The Key

Kevin McCormack

BUSINESS PROCESS ORIENTATION

Gaining the E-Business Competitive Advantage

BUSINESS PROCESS ORIENTATION

Gaining the E-Business Competitive Advantage

Kevin P. McCormack

William C. Johnson

S_L^t

St. Lucie Press

Boca Raton • London

New York • Washington, D.C.

Library of Congress Cataloging-in-Publication Data

McCormack, Kevin P.
  Business process orientation: gaining the e-business competitive
advantage / by Kevin P. McCormack and William C. Johnson.
    p. cm.
    ISBN 1-57444-294-5
  1. Industrial management—Data procesing. 2. Management information
systems. 3. Business enterprises— Automation. 4. Manufacturing
processes— Automation. 5. Marketing—Management—Data processing.
  I. Johnson, William C.   II. Title.
HD30.2 .M39 2000
658.4 — dc21

00-011197

Visit the CRC Press Web site at www.crcpress.com

DEDICATION

This book is dedicated to Susan. Her insights and perspectives have been invaluable both for this book and for my life. Her innate process orientation and system thinking has been my inspiration. She is the key competitive advantage in my life.

—**Kevin McCormack**

To my mother, whose selfless and sacrificial love over the years has been a constant source of encouragement and support.

—**Bill Johnson**

PREFACE

The old ways of conducting business are out: pushing costs and compensating quality in order to achieve the lowest possible price. A new paradigm is emerging with the integration of business partners and the focus on the core processes, according to Bernard Teiling, assistant vice president of Business Process Integration at Nestlé S.A.

The hallmarks of a great business model include high customer relevance, internally consistent decisions about scope and value chain activities performed, value capture mechanism, a source of differentiation and strategic control and a sound operational system and processes that are carefully designed to support the company's business model.[1] George Day, the Geoffrey T. Boisi Professor of Marketing at the Wharton School, suggests that key processes must be internally integrated and externally aligned with the corresponding processes of the firm's customers.[2]

Beginning with the outcomes of processes, reconfiguring internal processes based on changing customer requirements can help managers identify a different value chain, leading to a competitive advantage. To succeed in the future, corporations will have to weave their key business processes into hard-to-imitate strategic capabilities that distinguish them from their competitors in the eyes of customers. This is the very premise of our book. We believe that corporate survival in the Internet economy will depend both on the effectiveness of internal processes and their integration with supply chain customers. Supply chain management will serve as the coordinating mechanism for process integration among supply chain partners. Competitors can match individual processes or activities but cannot match the integration or "fit" of these activities.

Companies today are integrating their processes across the supply chain using networks, shared databases, the Internet, and extranets in order to quickly share information about customer requirements, production, delivery schedules, etc. Utilizing these connective technologies means that information is now available to the entire supply chain almost simultaneously.

Processes, as like never before, are now considered strategic assets. Witness how some dot-com firms like Amazon.com are protecting their business processes through patents, such as their one-click ordering and their Internet customer-based referral system (what Amazon calls "affiliates"). In fact, Amazon recently brought a court injunction against Barnes & Noble for that company to drop its own one-click feature.

Business Process Orientation: Gaining the E-Business Competitive Advantage was written to help business practitioners and academics understand the impact well-defined and carefully integrated processes have on organizational performance. The bulk of our insights and conclusions are drawn from actual research conducted among consumer, business-to-business, and services-based companies. Our research has demonstrated that adopting a business process orientation (BPO) has a positive impact on both the organizational culture and business performance.

Our book is organized into three sections. The first part of the book consists of nine chapters, beginning with an introduction and history of processes and process orientation (Chapters 1 and 2). Next, we present our research model and explain how the various measures of BPO were developed and tested (Chapter 3). Chapter 4 discusses our research model and presents the results of our field research. Chapters 5 through 7 administer the BPO measures in order to "benchmark" organizations' process orientation. Chapter 5 presents the BPO Maturity Model and explains the various stages of the model. Chapters 6 and 7 report research data collected using the BPO measure on two large manufacturing and service businesses and benchmark their progress based on the BPO Maturity Model. Chapter 8 discusses how a business process orientation affects supply chain management, utilizing a cross-industry study. Finally, based on the stage in the BPO Maturity Model, Chapter 9 provides a "prescription" of how to implement process initiatives to create superior value for the organization.

The second section of *Business Process Orientation: Gaining the E-Business Competitive Advantage* offers four current cases that provide hands-on examples of how process design and improvement create superior value and a sustained competitive advantage. Time Insurance and ABIG are primarily services-based organizations that have adapted their processes based on changing customer requirements. New South is a large, private lumber manufacturer whose story illustrates how changing manufacturing processes also involves changing the corporate culture. Finally, the Boston Market case shows how a change in business strategy can affect process effectiveness and, in this case, process flow.

The last section of the book contains the Appendices, which include the BPO measurements used both for individual companies' BPO and supply chain practices. We also included the statistical findings to supply more detail to the research results presented in Chapter 4.

Finally, you will note that our book cover has a Yin and Yang symbol. Incorporated within this is a hierarchical symbol to represent the vertical or functional orientation and a picture of people running toward the customer to represent the horizontal or business process orientation. These two conditions, as with the Yin and Yang symbol within which they are incorporated, are opposite and complementary and both must be present in healthy organizations. By balancing an organization's functional and horizontal orientation and maintaining that balance, leaders can tap into an energy reservoir that has been unavailable until now. We believe the higher levels of BPO will provide the balance needed between the vertical (functional hierarchy) and the horizontal (process). This balance is critical to the short- and long-term health of an organization. The illustration used on the cover of this book was designed to communicate this idea. We hope you enjoy reading the book and we welcome your comments. Feel free to contact either Kevin McCormack at 1-205-733-2096 or KMccorm241@aol.com or Bill Johnson at 1-800-672-7223 (ext. 5109) or billyboy@huizenga.nova.edu. You may also try our Website at www.bporientation.com.

Notes

1 Slywotzky, A., Morrison, D., Moser, T., Mundt, K., and Quella, J., *Profit Patterns*, New York, Times Business Random House, 1999.
2 Day, G., Managing market relationships, *Acad. of Mark. Sci. J.*, Winter 2000.

THE AUTHORS

Dr. Kevin McCormack has more than 25 years of business leadership and consulting experience in the areas of strategy, business process engineering, reengineering, change management, supply chain improvement, organizational design, and information technology implementation. His experience covers many national and international industry segments and a broad range of business processes. He has been a member of, or has successfully conducted engagements  with, several government agencies and major companies in the food, forest products, pharmaceutical, chemical, consumer products, high tech, and plastics industries. His clients have included Kraft, Philip Morris, CPC International, Cargill, Texas Instruments, Phillips Petroleum, Columbia Forest Products, Dow Chemical, Warner–Lambert, Standard Charter Bank, Microsoft, Tektronix, Borden Chemical, California Public Employees Retirement System, PepsiCo, and several state governments.

Dr. McCormack has held leadership positions in the food, beverage, chemical, consumer products, and information technology industries in the United States and in Europe. Dr. McCormack holds undergraduate degrees in Chemistry and Engineering, an MBA, and a DBA. He has taught Information Technology and Operations Management courses at the graduate and undergraduate levels in the United States and in Europe. Dr. McCormack's area of research is Business Process Orientation and its impact on business performance and IT investments.

Dr. McCormack is a member of the American Society for Quality (ASQC), the Supply Chain Council, the American Marketing Association (AMA), the American Production and Inventory Control Society (APICS), Council of Logistics Management (CLM), the Institute for Operations Research and the Management Sciences (INFORMS) and the Institute for Business Forecasting (IBF).

William Johnson is professor of marketing at the School of Business and Entrepreneurship, Nova Southeastern University. He teaches marketing courses at both the masters' and doctoral levels. Dr. Johnson has consulted with the soft drink, healthcare, telecommunications, cosmetic, and industrial chemical industries. He has worked with a variety of small businesses in Broward County in dealing with their marketing problems.

Dr. Johnson received his Ph.D. in Business from Arizona State University in 1985. He has taught in higher education for over 15 years. He has published in such journals as *The Journal of Applied Management and Entrepreneurship, Journal of Business and Industrial Marketing, Computers and Industrial Engineering International Journal, Marketing Education Review, The Journal of Marketing in Higher Education, Marketing News, International Business Chronicle, Arizona Business Education Journal, The Marketing Connection, Industrial Engineering International Journal,* and *Beverage World.* He has co-authored two textbooks, *Total Quality in Marketing* and *Designing and Delivering Superior Customer Value: Concepts, Cases and Applications,* published by St. Lucie Press, Boca Raton, FL. Dr. Johnson has had experience in international education, presenting seminars to business professionals from Brazil, Taiwan, Thailand, and Indonesia.

TABLE OF CONTENTS

1

INTRODUCTION

Recently, General Electric CEO John F. Welch, Jr. ordered a move to e-processes, applying business-to-business technology everywhere. For example, at GE Information Services, employees use a system called Trading Partner Network Register to order office supplies from pre-qualified vendors over the Internet. By GE estimates, making purchases offline can cost between $50 and $200 per transaction, while online costs amount to only about $1 per transaction.

IBM conducted a wholesale review of its processes a few years ago. Realizing that its large corporate customers were increasingly operating on a global basis, IBM knew it would have to standardize its operations worldwide. It would have to institute a set of common processes for order fulfillment, product development, and so forth to replace the diverse processes that were then being used in different parts of the world and in different product groups. IBM even changed its management structure, assigning each major process to a member of its senior-most executive body. Further, each process was assigned an owner, referred to as a business process executive, who was given responsibility for designing and deploying the process. Each of IBM's business units is now expected to follow processes designed by their business process executives. Shifting organizational power away from units and toward processes has helped IBM standardize its processes around the world. The benefits have been startling, with a 75% reduction in the average time to market for new products, a sharp upswing in on-time deliveries and customer satisfaction, and cost savings in excess of $9 billion.

Giant retail broker firms like Merrill Lynch and PaineWebber for years have excelled at four business processes crucial to overall business success: client management, information delivery, portfolio modeling, and operational statistics. However, with the Internet fast becoming the preferred channel among investors, online trading has emerged as a fifth critical process. PaineWebber and Merrill Lynch, with their fat brokerage fees ranging in the hundreds of dollars, reluctantly began shifting some of their business to the Internet.

Federal Express recently announced plans to launch an online service that will enable the delivery company's business customers review and pay invoices over the Internet. FedEx, a unit of FDX Corporation, said the electronic bill-presentment and bill-payment service, called Invoice online, will allow customers to schedule payments as many as 15 days in advance. A second, and arguably more ambitious process improvement effort, involves FDX trying to recast itself as a major provider of supply chain management systems that threaten the company's very existence. FDX plans to design a network that can supplant a company's inefficient stream of faxes and phone calls with digital exchanges of information about demand, factory schedules, and availability of materials. Such systems would select the most logical, most economical type of transport, whether air, land, or sea, for delivering packages on time. FDX would then coordinate customs clearances around the world and minimize the amount of time any item sits in a warehouse along the way.

There is increasing evidence from these and other successful companies that a superior competitive advantage results from a combination of the organization's assets (brand image and marketing capabilities) and skills (e.g., innovation), which, when applied advantageously to business processes, results in superior customer value. According to Mroz, "In the information economy of the twenty-first century, corporate survival will depend on the effectiveness of the corporation's innate business processes...corporations will be defined not so much by their industry or products, but by the nature of their processes."[1]

Today, traditional value chains are under threat as the processes that underpin business relationships continue to evolve, where knowledge creation and innovation are replacing physical processes as the critical value-adding activities. The Internet in particular is forcing companies to reconfigure their internal value chains, especially in the buying and selling of goods and services. A recent worldwide survey of 500 large companies carried out jointly by Economist Intelligence Unit and Booz-Allen & Hamilton, found that more than 90% of top managers believe the Internet will transform or significantly impact the global marketplace by 2001.

Corporate purchasing is easily the most attractive candidate for e-commerce. Deloitte Consulting LLC estimates that 91% of U.S. businesses will do their purchasing on the Net by the end of next year, whereas some 31% do so now. Nowhere is this change more apparent than the automobile industry, where Ford Motor Company and General Motors recently unveiled plans to go online with their massive purchasing systems, which each year acquire $80 billion and $87 billion, respectively, in goods and services. Ford is partnering with Oracle Corp. to create *AutoXchange*, a purchasing system that will use an online auction to fill orders. GM is teaming up with Honda to offer *TradeXchange*, their own web-based system which GM hopes will streamline their purchasing process and allow buyers to aggregate their purchases electronically.

We have already seen during this nascent Internet era that well-designed processes can make a huge difference in the success or failure of consumer e-commerce ventures. During the recent 1999 Christmas shopping season, many e-tailers came under heavy criticism for failing to deliver toys on time for Christmas. Countless shoppers were left empty-handed not only because of late deliveries, but also because products were out of stock, sites were down and customer service was almost nonexistent. Toysrus.com had an especially stormy Christmas season. Toys "Я" Us Inc.'s Internet division is being sued by a customer who claims the company failed to deliver thousands of Christmas toys on time. With online sales of $39 million from November 1 to December 25, toysrus.com received far more orders than expected and was forced to turn away a number of customers in November.

Online shoppers are sending a clear message: e-tailers who fail to improve their delivery and service responsiveness risk losing future patronage. Efficient order fulfillment is not the only concern of Web shoppers. Although they like the convenience of Web shopping, consumers are becoming increasingly frustrated with the other elements of the buying process, such as the difficulty of entering information. According to *The New York Times*, consumers bail out of online transactions before they are completed 30 to 60% of the time.

Building an attractive Website is merely a starting point. E-commerce companies, both consumer and business-to-business, need to pay careful attention to the back-end processes that generate orders which are processed and delivered in a timely fashion. We view a business process orientation (BPO) as a way for firms to get closer to their customers by improving organizational performance and competitiveness. Whether conducting consumer or business-to-business e-commerce, a BPO is critical for designing processes which translate into superior customer value. To succeed in the year 2000 and beyond, corporations will have to weave their key business processes into hard-to-imitate strategic capabilities that distinguish them from their competitors in the eyes of customers. Process mastery will be a key factor in achieving a sustainable competitive advantage in the Internet economy. However, process mastery needs to be understood in the context of customer value, the subject of the next section.

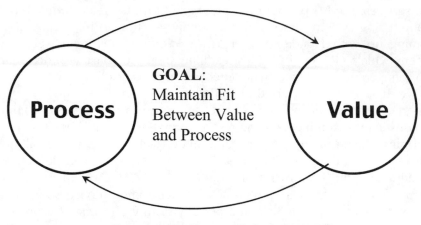

Process Delivers Value through:

• Quality

• Cost Reduction

• Flexibility

Figure 1.1 The Link between Process and Value

PROCESS AND VALUE

Processes and value chains are evolving rapidly as companies outsource non-core activities and capabilities. A critical decision by business managers today is what and how to deliver the firm's *core processes*. This decision should be made based on a simple litmus test: will the process lead to superior customer value? As Figure 1.1 shows, the goal of the organization is to maintain a fit between value and processes. Successful organizations recognize that value and process are "seamless" in the eyes of their customers. Ford recently announced that it was organizing its dealer service area around four key processes that create customer satisfaction. Sears, Roebuck & Co. and French retailer Carrefour recently announced an Internet retail exchange to handle the $80 billion they spend annually on supplies. They have even invited other retailers to join. What prompted these organizations to change their processes? In short, they desired to better serve their customers and in the process deliver greater value.

KEY ORGANIZATIONAL PROCESSES

Before discussing key organizational processes, let us define what we mean by "process." A process is a specific group of activities and subordinate tasks which results in the performance of a service that is of value. Business process design involves the identification and sequencing of work activities, tasks, resources, decisions, and responsibilities across time and place, with a beginning and an end, along with clearly identified inputs and outputs. Processes must be able to be tracked as well, using cost, time, output quality, and satisfaction measurements. Businesses need to continually monitor, review, alter, and streamline processes in order to remain competitive. A process view of the organization differs from the traditional functional view, as presented in Table 1.1.

In fact, organizations that view themselves as a collection of processes that must be understood, managed, and improved are most likely to achieve this end. Thus, firms need to shift their focus from managing departments to managing processes. Most organizations today are aligned along departmental lines, that is, warehouse, customer service, purchasing, etc. This structure is inefficient and costly. The focus is typically on whose fault it is and not on how we can satisfy the customer. Customer needs are not met by departments but by processes that cut across departmental lines.

So why don't businesses take a process view of their organizations? While many companies have integrated their core processes, combining related activities and cutting out ones that don't add value, but only a few have fundamentally changed the way they manage their organizations. The power in most companies still resides in vertical units sometimes focused on regions, sometimes on products, and sometimes on functions. These fiefdoms still jealously guard their turf, their people, and their

Table 1.1 Process View vs Traditional Functional View

Process View	Functional View
Emphasis on improving "how work is done"	Which products or services are delivered
Cross-functional coordination, teamwork stressed	Frequent "hand-offs" among functions which remain largely uncoordinated
"Systems view," i.e., entire process is managed	Pieces of the process are managed
Customer orientation	Internal/company orientation

resources. The combination of integrated processes and fragmented organizations has created a form of cognitive dissonance in many businesses: the horizontal processes pull people in one direction; the traditional vertical management systems pull them in another. The confusion and conflict that ensue ultimately undermine business performance.

Processes are not simply obscure, back-room operations of the service concern, but instead an integral part of delivering the value proposition. We maintain that processes and service are inseparable, that is, the process *is the service*. An effective process is results driven, deriving its form from customer requirements, such as how and when customers want to do business with you. Market-oriented companies ensure that the service encounter is positive by asking: how can we make our customers' life easier? GE asked that question and came up with the idea of GE's Answer Center, a fully staffed customer call center that operates 24 hours a day offering repair tips and helping owners of GE appliances with their problems. We recommend that managers first take a "big picture" view of their companies by looking at key processes in relationship to the marketing cycle.

Figure 1.2 shows the marketing cycle and how it relates to business processes and process indicators. You will note that the various market constituents such as customers, suppliers, and publics determine how and to what extent the marketing cycle elements are performed. Customers, in particular, determine the composition and nature of the marketing cycle and the subsequent core processes that are required to support these selected marketing cycle functions. For example, the customer service *process* is performed as part of the service management function of the marketing cycle. Customer service activities would include, but are not limited to, such activities as tracking and trending customer complaints, recovery from customer service failures, and establishing customer service standards. The process indicators represent the "metrics" for measuring the core processes. One of the process indicators for the customer service process is gauging customer satisfaction levels. Ford tracks customer retention as part of its service management process and has found that each additional percentage point in customer retention rates is worth $100 million in profits. It should also be pointed out that a synergy exists within the marketing cycle elements. That is, process breakdown in one area, such as logistics, affects other areas such as distribution.

But just as important as having smooth, efficient processes with appropriate metrics is being able to redesign those processes as market conditions change. From order fulfillment to customer service to procurement, operating processes are rarely fixed any more. They must change their shape as markets change, as new technologies become available, and as new competitors arrive. IBM redesigned most of its processes over the last few years to make them compatible with CEO Gerstner's web-centric strategy. The next section considers some critical steps in assessing process effectiveness.

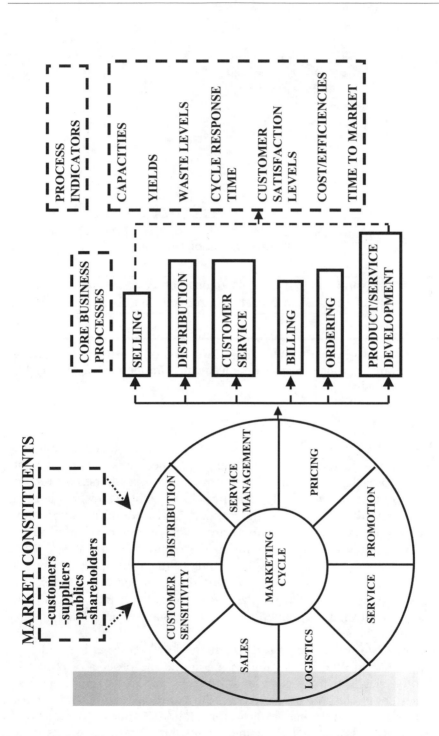

Figure 1.2. The Marketing Cycle and Process Model

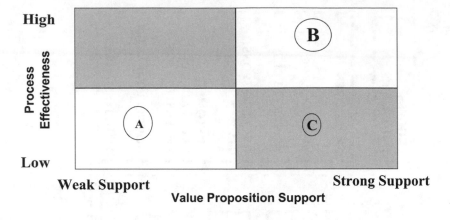

Figure 1.3. Process Support of Value Proposition (Source: J. Feather, "Using Value Analysis to Target Customer Service Improvements," *Business Week,* January 17, 2000, with permission)

Assessing Process Effectiveness

It is necessary to assess process effectiveness before implementing process change or improvement. We suggest that companies follow a fairly straightforward approach to assessing their current processes. First, we recommend that companies *define* what it is they do and where they are planning to go. In other words, what is the company's vision and mission? Some questions that need to be answered are: What is our core purpose for being? What is the overall direction that the company wants to go? What opportunities can and should be pursued? Is the value proposition still relevant? The *Business Process Assessment Tool* included in Appendix B is extremely helpful for diagnosing the present process readiness of a company and we strongly recommend its use as a "starting point" in assessing process effectiveness.

The next step is to *understand* what the key processes are and how they are related to the firm's value proposition. The process or processes need to be clearly defined, including the steps that make up the process. Processes should also be assessed according to their efficacy and congruence with the firm's value proposition. John Feather, a partner with Corporate Renaissance, a management consulting group, suggests using

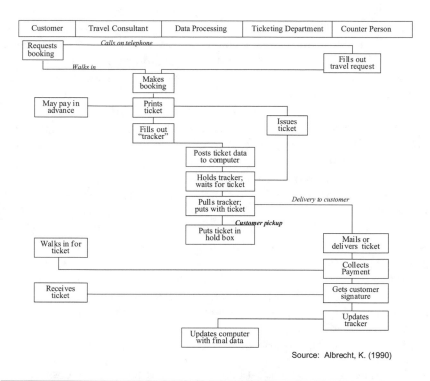

Customer	Travel Consultant	Data Processing	Ticketing Department	Counter Person

Requests booking

Calls on telephone

Fills out travel request

Walks in

Makes booking

May pay in advance

Prints ticket

Fills out "tracker"

Issues ticket

Posts ticket data to computer

Holds tracker; waits for ticket

Pulls tracker; puts with ticket

Delivery to customer

Customer pickup

Puts ticket in hold box

Walks in for ticket

Mails or delivers ticket

Collects Payment

Receives ticket

Gets customer signature

Updates tracker

Updates computer with final data

Source: Albrecht, K. (1990)

Figure 1.4. Service Flow Diagram (Source: Karl Albrecht, *Service Within*, Homewood, IL, Irwin, 1990, reproduced with permission of the McGraw-Hill Companies)

a grid similar to Figure 1.3 to ensure that processes are aligned with the firm's value proposition.[3]

Further, it is important to conceptualize not only which steps are performed but also the timing and sequencing of relationships in the process. Blueprinting the steps of the process can help visualize the actual steps in the process as well as the process flow. A process flow diagram like Figure 1.4 should be used to help identify "fail points," or steps in the process that are likely to go wrong. Time Insurance developed a process map that charted the flow of work required to issue a new policy, described in terms of "blocks of activity" (see Appendix A, Time Insurance Case).

It is next to impossible to assess processes without well-defined *standards*. Policies, procedures, and routines are needed to enable employees to perform their jobs effectively and efficiently. Compliance or certification programs such as ISO 9000 help support this effort. Such standards provide a means of accountability that a company's processes work as stated and documented.

Process management, particularly process improvement, requires proper *measures*, a fourth step in assessing process effectiveness. According to Tenner and DeToro, there are three ways in which to measure performance: **process measures**, which define activities, variables and operation of the work process itself; **output measures**, which define specific characteristics, features, values, and attributes of each product or service; and **outcome measures**, which measure the impact of the process on the customer and what the customer does with the product or service (customer satisfaction measures are often used here to evaluate outcome measures).[4] Table 1.2 provides examples of some core processes and appropriate output measures.

After using appropriate process assessment measures, the final step in process assessment involves process *improvement*. Here processes need to be either fine-tuned or completely reengineered, based on whether they are "out of tolerance." Process quality tools such as Pareto diagrams and control charts are well suited to provide employees with feedback on job and process performance. The decision whether to modify or completely reengineer core processes should be informed by customer requirements. For example, the prestigious Karolinska Hospital in Stockholm, Sweden, reorganized its key processes around patient flow, instead of allowing the patient to be bounced from department to department. Some of the more common approaches to process improvement include:

- Eliminate tasks that have been determined to be unnecessary
- Simplify the work by eliminating all non-productive elements of a task
- Combine tasks

Table 1.2 Marketing Cycle Functions and Output Measures

Marketing Cycle Function	Core Marketing Process	Outcome Measure
Distribution	Delivery	% On-Time Deliveries
Promotion	Media Selection	Cost per Thousand
Logistics	Order Fulfillment; Billing	Transaction Time Billing Accuracy
Sales	Prospecting; Complaints Handled	Leads; Conversions; Complaint Resolution
Product Management	Product Development Process	Time-to-Market; New Product Success Rates

- Change the sequence to improve speed
- Perform activities simultaneously

Savin Corporation, a larger copier company, conducted a careful study and found that callbacks (callbacks are where technicians are sent out on service calls) were related to deficiencies in the training process. Pareto diagrams were prepared depicting those service engineers responsible for the largest number of callbacks. It was determined that training just five engineers would reduce callbacks by 19%. In most cases, the people who perform the specific processes that are under study are the ones most capable of determining how to improve or simplify the process.

The focus of this book is how to practice a business process orientation (BPO), that is, designing business operations and processes that are value creating. We begin by reviewing the history of business process orientation and examining the early contributors to BPO. We then discuss how to define and measure BPO, reporting research on how to evaluate BPO. We will then examine how BPO leads to superior organizational performance, again reporting our own research results. Next, we introduce benchmarking, using our BPO Maturity Model to help firms determine where they are and where they need to be. The last section of the book explains how to apply BPO to manufacturing and service operations, using BPO to guide key process areas in the supply chain. Finally, we conclude by offering prescriptive approaches to implementing and evaluating BPO.

SUMMARY

Business today is driven more and more by speed and efficiency. Companies that get their products/services to market first, develop seamless links with their suppliers, and fill orders when promised will be the survivors in the new economy. A full understanding of process relative to customer requirements will be key to achieving a competitive advantage in the brave new world of e-commerce. Companies now use Internet links to collaborate with trading partners on product development, logistics, and sales efforts, resulting in a campaign that is much more responsive to evolving customer needs. A process orientation helps companies think about how their activities and tasks either add or subtract customer value. Creating greater customer value through process orientation requires a disciplined approach, beginning with aligning core business processes with the firm's value proposition. Standards are also critical for any meaningful process improvement to take place.

PROCESS IN FOCUS[5]

When Canadian Pacific Hotels set out to gain a competitive advantage through closer relations with business travelers, it realized that it needed to realign its organization around team-based processes that cut across functions. Canadian Pacific Hotels, with 27 hotels in the quality tier across Canada, has been proficient with conventions, corporate meetings, and group travel but wanted to excel with business travelers. This is a notoriously demanding and difficult group to serve, but also a lucrative group much coveted by all other hotel chains. When conducting in-depth research on this important market segment, Canadian Pacific Hotels found that frequent guest programs had little appeal, because these road warriors preferred airline mileage. Travelers also appreciated beyond-the-call-of-duty efforts to rectify problems when they happened. Above all, travelers wanted recognition of their individual preferences and lots of flexibility on when to arrive and check out.

Canadian Pacific Hotels responded by committing to customers in its frequent-guest club to make extraordinary efforts to always satisfy preferences for type of bed, location in hotel (high or low), and all the other amenities. Delivering on this promise proved remarkably difficult. Canadian Pacific Hotels began by mapping each step of the "guest experience" from check-in and parking valet to checkout and setting a standard of performance for each activity; then determining what had to be done to deliver on the commitment to personalized service. What services should be offered? What processes were needed? What did the staff need to do or learn to make the process work flawlessly?

A major challenge was Canadian Pacific Hotels' historic bias toward handling large tour groups. The skills and processes at hand were not the ones needed to satisfy individual executives who did not want to be asked about their needs every time they checked in. Even small enhancements such as free local calls or gift shop discounts required significant changes in information systems. The management structure was changed so each hotel had a champion with broad, cross-functional authority to ensure the hotel lived up to its ambitious commitment. Finally, further systems and incentives were put in place to ensure that every property was in compliance and performance was meeting or exceeding the standards. In a business that demands constant attention to innumerable details, no single factor determines whether a customer will be loyal. It is the sum of many elements that makes the difference and the market rewards the effort. In 1996, Canadian Pacific Hotel's share of Canadian business travel jumped by 16%, although the total market was up just 3%, and Canadian Pacific Hotels added no new properties. By all measures, Canadian Pacific Hotels is winning greater loyalty from its target segment.

NOTES

[1] Mroz, R., Unifying marketing: The synchronous marketing process, *Industrial Mark. Manage.*, Vol. 27, 1998.

[2] Cohn, L., Brady, D., and Welch, D., B2B: The hottest net bet yet, *Business Week*, January 17, 2000.

[3] Feather, J., Using value analysis to target customer service improvements, *IIE Solutions*, May, 1998, pp. 33–39.

[4] Tenner, A. and DeToro, I. *Total Quality Management*, Reading, MA, Addison–Wesley Publishing, 1992, p. 44.

[5] Adapted from Day, G. Managing market relationships, *Acad. of Mark. Sci. J.,* Winter 2000.

2

HISTORY OF BUSINESS PROCESS ORIENTATION

This chapter reviews the evolution of business process orientation (BPO), beginning with the concept of functional orientation that began at the turn of the century through the Total Quality Management (TQM) phase of the 1980s, the reengineering craze of the 1990s, and the current e-business frenzy. The introduction of foundation process concepts and contributions by Edward Deming, Michael Porter, Peter Drucker, and others are discussed, as is the process thinking introduced by the Japanese.

The orientation of a firm and the base point of reference for the people in the firm are critical aspects of all the business drivers. This "way of looking at the world" drives strategy, decision-making, investments, and selection of employees and leaders. A study of U.K. manufacturers attempting to examine orientations in these firms identified the following types and descriptions of orientations.[1]

Production: Concentrate on reducing costs, achieving high production efficiency and productivity and increasing production capacity.

Product: Make products with good quality and features, improve them over time, and then try to sell them.

Selling: Concentrate on promoting and selling what we can make.

Market: Identify changing customer wants and develop products to serve them better than competitors.

Competitor: Identify the closest rivals, learn their strengths and weaknesses, forecast their behavior, and develop marketing strategies to capitalize on their weaknesses.

BPO was significantly missing from this list. Why? Did this orientation not exist or was it just not defined enough to measure and talk about?

Most of what has been written regarding BPO during the last two decades is in the form of success stories concerning new forms of organizations. Although empirical evidence is lacking, several examples of these new forms have emerged during this period that have been presented as high performance, process-oriented organizations that are needed to compete in the future. Authors such as Deming, Porter, Davenport, Short, Hammer, Byrne, Imai, Drucker, Rummler–Brache, and Melan have all defined what they view as the new model of the organization. Developing this model requires a new approach and a new way of thinking about the organization, which will result in dramatic business performance improvements. This new way of thinking or viewing the organization has been generally described as *business process orientation* or BPO.

During the 1980s, Michael Porter introduced the concepts of interoperability across the value chain and horizontal organization as major strategic issues within firms.[2] Edward Deming developed the "Deming Flow Diagram" depicting the horizontal connections across a firm, from the customer to the supplier, as a process that could be measured and improved like any other process.[3] In 1990, two researchers, Thomas Davenport and James Short, proposed that a process orientation in an organization was a key component for success.[4] In 1993, Michael Hammer, who led the reengineering craze of this decade, also presented the business process orientation concept as an essential ingredient of a successful "reengineering" effort. Hammer described the development of a customer-focused, strategic business process-based organization enabled by rethinking the assumptions in a process-oriented way and utilizing information technology as a key enabler.[5] Dr. Hammer offered reengineering as a strategy to overcome the problematic cross-functional activities that present major performance issues to firms. The apparent conflict between a functional focus (whom I report to) vs. a horizontal focus (whom I provide value to) is offered by Hammer as being brought back in balance by adding a business process orientation to the organization.

As the "e-craze" of this decade (e-business, e-commerce, e-supply chain) replaces the reengineering craze of the 1990s, business process performance and the horizontal nature of e-corporations have risen to new levels of importance. Corporations are extending outside their legal boundaries as a normal way of organizing. Partnering, functional outsourcing, business process outsourcing, alliances, and joint ventures are yesterday's requirements for success. Today's success depends on new e-forms of horizontal and vertical "virtual integration" that are appearing each day. Business process orientation is not simply a way to organize but an imperative for survival.

The remainder of this chapter presents the key contributions to the history of business process orientation and the imperatives for the e-corporation.

FUNCTIONAL ORIENTATION: 200 YEARS AND COUNTING

In 1776, Adam Smith described the concept that industrial work should be broken into its simplest tasks. This idea became the basic organization model of business for almost 200 years. The modern business enterprise has gone through only two major evolutions since the Civil War in the United States.[6] Around the turn of the century, management came to be viewed as work in its own right. Up until that time, management was indistinguishable from ownership. J.P. Morgan, Andrew Carnegie, and John D. Rockefeller began the restructuring of the railroads and American industry using the basic principles of Adam Smith and the new concept of management work or hierarchy. Twenty years later, Pierre S. DuPont began the second evolution by restructuring the family business into the modern corporation. Alfred Sloan redesigned General Motors and further defined this business model. This institutionalized command and control, centralization, central staffs, the concept of personnel management, budgets and controls. This model is our tightly defined, tightly controlled, functionally centered organization model of today.

Business performance, as defined by return on assets (ROA), was realized with this model through the leverages of size and division of labor. This allowed organizations to maintain highly paid, scarce skills, as well as effectively gather and deploy natural resources and labor, the two major factors in the success of enterprises of the time. The hierarchy of skilled managers was necessary to coordinate the functional activities, manage the information flow, and interface with the other functions in the organization. The better the focus and coordination of the company resources, the more profitable the business.

The functional view of the organization is best described by the organization chart (see Figure 2.1).

This chart shows which people have been grouped together for operating efficiency and illustrates reporting relationships. What is not shown is the customer and the what, why, and how of the business. In functionally centered organizations, hand-offs between functions are frequently uncoordinated. The greatest opportunity for performance improvements lies in the functional interfaces, or the points where the "baton" is being passed from one function to another.

Too often, what is being managed is power and authority, not the activities that bring value to the customer.

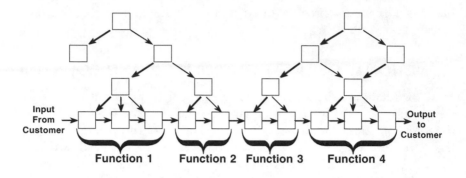

Figure 2.1 The Typical Organization Chart

BUSINESS PROCESS ORIENTATION IN THE 80s: BPO FOUNDATIONS

The concept of improving these functional interfaces by "viewing" the business differently is evident in Edward Deming's philosophy, captured by "The Deming Flow Diagram" (see Figure 2.2).[7]

The flow diagram takes a business process orientation and describes a business as a continuous process connected on one end to the supplier and on the other end to the customer. A feedback loop of design and redesign of the product is also shown as connected to both customers and suppliers. Deming's 14 points and elimination of the seven diseases describe the strategies for optimization of the flow diagram and therefore the creation of superior customer value and superior profitability.

In 1985, Michael Porter introduced the "value chain" concept as a systematic way of examining all the activities a firm performs and how they interact to provide competitive advantage (see Figure 2.3). This chain is composed of "strategically relevant activities" that create value for a firm's buyers. Competitive advantage comes from the value a firm is able to create for its buyers which exceeds the firm's cost of creating it.

A firm gains competitive advantage by performing these strategically important activities more cheaply or better than competitors. According to Porter, a firm is profitable if the value it commands exceeds the costs involved in creating the product.

A major way to develop competitive advantage in this value chain is described by Porter as managing linkages. Linkages are relationships between the way one value activity is performed and the cost of performance of another. Optimization and coordination approaches to these linkages can lead to competitive advantage. The ability to coordinate linkages often reduces cost or enhances differentiation. This recognition

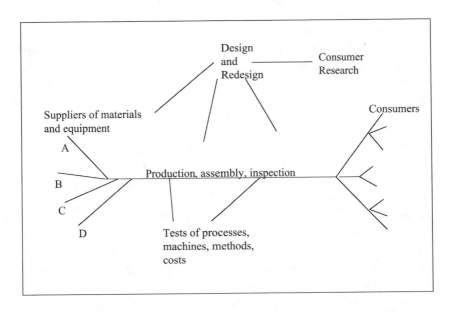

Figure 2.2 The Deming Flow Diagram (Adapted from M. Walton, *The Deming Method*, New York, Perigree Books, 1986, p. 28)

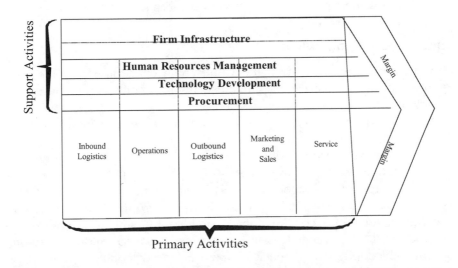

Figure 2.3 The Generic Value Chain (Adapted from M.E. Porter, *Competitive Advantage: Creating and Sustaining Superior Performance*, New York, The Free Press, 1985, p. 37)

of the importance of linkages, according to Porter, has been strongly influenced by Japanese management practices. The ability to recognize and manage linkages that often cut across conventional organizational lines can yield a competitive advantage. The linkages between supplier and customer value chains can also be a source of competitive advantage.

The organizational structure often defines the linkages in a value chain. Integrating mechanisms must be established to ensure that the required coordination takes place. Information is essential for the optimization of these linkages and is rarely collected or connected throughout the chain. Porter suggested that a firm might be able to design an organization structure that corresponds to the value chain and thus improve a firm's ability to create and sustain competitive advantage through coordination, minimization, and optimization of linkages.

Michael Porter's value chain is a method to define a business in a customer-focused, strategic-process-oriented way. Porter does not go into the details of coordination and optimization of linkages but suggests that a new organizational model can have a major impact on a firm's performance. It is clear that the closer the organizational structure is to the way the strategic processes are organized, the more effective it can be in providing value. According to Porter, this value will lead to competitive advantage and profitability.

The Porter value chain and the suggestion that a firm organized around this structure can gain a strategic competitive advantage positioned the concept of business process orientation firmly as a key competitive strategy.

The Japanese Contribution

Shortly after Porter introduced the value chain concept, a popular management principle, kaizen, the Japanese management principle that has reportedly given many companies a competitive advantage, was introduced.[8] This principle added a new dimension to the orientation of an organization.

Masaaki Imai, a leading Tokyo-based management consultant, unequivocally stated at that time that "kaizen strategy is the single most important concept in Japanese management — the key to competitive success" (Imai, 1986). Kaizen, as explained by Imai, is the overriding concept behind good management: a combination of philosophy, strategy, organization methods, and tools needed to compete successfully today and in the future.

The philosophy component of kaizen is one of continuous improvement of everything, every day, and involving everyone. This, said Imai, is the unifying thread running through the philosophy, systems, and problem-solving tools developed in Japan over the last 30 years.

The strategy consists of (a) recognizing that there are problems and establishing a corporate culture in which everyone can freely admit these problems; (b) taking a systematic and collaborative approach to cross-functional problem-solving; (c) a customer-driven improvement strategy; (d) significant commitment and leadership of kaizen from top management; (e) an emphasis on process and a process-oriented way of thinking; and (f) a management system that acknowledges people's process-oriented efforts for improvement.

The kaizen tools consist of various approaches, methods, and techniques that analyze and organize the process and improvements efforts. Contributors include Deming, Juran, and many of the quality leaders. Statistics, systematic problem solving, charting, and teamwork are stressed in many of the kaizen tools.

Perhaps the major point stressed by Imai is that management must adopt a process-oriented way of thinking. Japan is described as a process-oriented and people-oriented society whereas the U.S. is described as a results-oriented society. In a results-oriented society, only results count. In a process-oriented society, improvement efforts count. Neither approach, taken by itself, is the "right" way as described by Imai. Results-oriented tends to focus only on the what, thus neglecting the how, while the process-oriented focuses on the how, neglecting the what. Both have demotivating and defocusing issues. Imai proposes a combination of the two, using the strengths of both. The implementation of this philosophy must also be embodied in the reward and recognition system of the organization. Imai proposes that the implementation of kaizen will lead to an organization with reduced conflict and improved connectedness across the departments of the firm.

The Information Society

In 1988, Peter Drucker foresaw the need for a new organization model: given the major shifts in the environment, the old organization model was obsolete and a major barrier to competitiveness.[9] Demographics, economics, society, and, above all, information technology, all demanded a shift to an "information-based organization." This model consists of an organization of knowledge specialists organized in task-force teams. Traditional departments will serve as guardians of standards, centers for training, and the source of specialists but they won't be where the work gets done. The task-focused teams will work on a "synchrony" of activities or processes that span the old organizational boundaries and end with the customer. A sequence of tasks with hand-offs between functional groups will not exist.

Even before the e-craze or before the Internet came into commercial use, Drucker foresaw that the availability of information would transform the organization structure into a flat organization of specialists working on task-focused teams. The layers of command and control managers will not be needed. Some centralized service staffs will still be needed but the need will shrink drastically. Drucker said that this will require greater self-discipline and an ever-greater emphasis on individual responsibility for relationships and communication. The workers in this organization cannot be told how to do their work, since they, not management, are the experts. They will require clear, simple common objectives that translate into particular actions. Leadership will focus the skill and knowledge of the individuals on the joint performance of the organization similar to an orchestra being lead by a conductor. Drucker's model appears to describe a process-oriented, customer-focused, team-based organization of empowered specialists held together by a common vision and goals.

As with the other models discussed thus far, the implication is that this will lead to a firm's success if the management challenges can be overcome. Removing the functions from the process eliminates the interoperability issues and linkages between functional groups. This organizational model's linkage coordination and optimization will, using Porter's and Deming's principles, lead to a significant competitive advantage. If a solution to the management and reward issues can be found, this model would be a significant advance in organizational technology that would lead to reduced conflict and improved connectedness in a firm.

Table 2.1 summarizes the views of the key authors reviewed who have proposed a new model leading to improved cross-functional interoperability and improved business performance.

BUSINESS PROCESS ORIENTATION IN THE 90s: TECHNOLOGY ENABLEMENT

Dr. Michael Hammer started the reengineering craze when he declared war on the old organizational model in 1990 with his article, "Reengineering Work: Don't Automate, Obliterate," published in the *Harvard Business Review*.[10] His premise was that the old model, built in the 19th century, was no longer relevant and something entirely different was needed. This new model would be accomplished by looking at fundamental processes of the business from a cross-functional perspective and enable a radical new way of operating, using information and organizational technology. The radical new processes would drive dramatic changes in jobs and organizational structures. This, in turn, would require radical changes in the management and measurement systems that would shape the values and beliefs of the organization. These values and beliefs of the organization

Table 2.1. Summary of New Model Views — The Foundations of BPO

	Deming	Imai	Porter	Drucker
Strategic Focus	Long-term focus on customer value Constancy of purpose	Customer focus	Strategic relevant activities	Information based
Leadership/ Management	Coaching	Problem-solving culture	Linkage Management	Joint perform-ance focus
Reward/ Recognition	Long term based on customer and team	Process and Results oriented	Integrating	Team and develop-ment based
Structure	Continuous process, teams, supplier partners	Cross-functional, supplier partnerships	Fits value chain, supplier/ customer links, integrating mechanisms	Customer-focused processes, teams, develop-ment groups
Philosophy	Continuous improve-ment, empower-ment, teams, training, and education	Continuous improve-ment, systematic collabor-ation, process thinking	Manage/ optimize links, customer focused	Process oriented, customer focused, specialist, development
Tools/ Techniques	Data tools, stats	Analyze, organize, improve (stats, charts, JIT)	Value chain analysis	
Information	N/A	N/A	Essential for optimized links, connected throughout chain	Basis of organization driver

would finally support and enable the radically new business processes by reflecting the important performance measures of the new process.

Hammer defined a business process as a collection of activities that takes one or more kinds of input and creates an output that is of value to the customer. A reengineered business is composed of strategic, customer-focused processes that start with the customer and emphasize outcome, not mechanisms. This is the heart of the enterprise; how a company creates value and represents the real work.

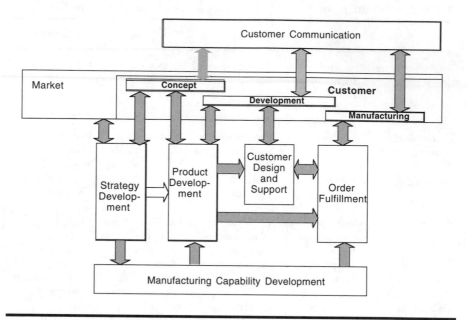

Figure 2.4 Texas Instruments High Level Business Process Map (Adapted from M. Hammer and J. Champy, *Reengineering the Corporation: A Manifesto for Business Revolution*, New York, Harper Business, 1993)

Process thinking is described as cross-functional, outcome-oriented, and essential to customer orientation, quality, flexibility, speed, service, and reengineering. A company is defined not by its products and services, but by its processes. Managing a business means managing its processes. These processes are classed as value adding, enabling, asset creating, and governing. Figure 2.4 is an example of a company, Texas Instruments Semiconductor Division, viewed as a process according to Dr. Hammer.

The construction of this map not only creates a process "view" of a business but it creates a process vocabulary that is essential for cooperation and coordination within the firm. This map makes visible the business processes that were invisible.

Hammer described the following changes that occur in the new process-oriented model.

1. Work units change from functional departments to process teams.
2. Jobs change from simple tasks to multi-dimensional work.
3. People's roles change from controlled to empowered.
4. Job preparation changes from training to education.

5. Focus of performance measures and compensation shifts from activity to results.
6. Advancement criteria change from performance to ability.
7. Values change from protective to productive.
8. Managers change from supervisors to coaches.
9. Organizations change from hierarchical to flat.
10. Executives change from scorekeepers to leaders.

Information technology enables the new organization to use the organizational technology components to build a high performance, customer-focused, empowered, flat, results-oriented, continuous improvement-oriented, and process-oriented organization. This organization model, according to Hammer, would result in dramatic increases in business performance and profitability.

Thomas Davenport, in his book *Process Innovation: Reengineering Work through Information Technology*, provided the foundation for this technology-oriented area of investigation by describing the needed revolutionary approach to information technology in business. This approach was new in how a business was viewed, structured, and improved.[11] Davenport suggested that business must be viewed as key processes, not in terms of functions, divisions, or products. One of Davenport's major propositions is that the adoption of a process view of the business with the application of innovation to key processes will result in major reductions in process cost, time, quality, flexibility, service levels, and other business objectives, thus leading to increased profitability.

The process view, according to Davenport, facilitates the implementation of cross-functional solutions and the willingness to search for process innovation, thus achieving a high degree of improvement in the management and coordination of functional interdependencies.

Davenport described having a process view, or a process orientation, as involving elements of structure, focus, measurement, ownership, and customers. A process itself was defined as "a specific ordering of work activities across time and place, with a beginning, an end, and clearly identified inputs and outputs a structure for action." The existing hierarchical structure is a "slice in time" view of responsibilities and reporting relationships. A process structure is a dynamic view of how an organization delivers value. Processes, unlike hierarchies, have cost, time, output quality, and customer satisfaction measurements. Process improvements can easily be measured. A process approach to business also implies a heavy emphasis on improving how work is done, in contrast to a focus on which specific products or services are delivered. In a process-oriented organization, investments are made in processes as well as products. The definition and structuring of processes themselves lend them to measurements and improvements in inputs and outputs. The consistency, variability, and freedom from defects can be defined and measured once the process is defined. This provides a focus and feedback loop that facilitates improvement.

Davenport's process approach implies adopting the customer's point of view. A measure of customer satisfaction with the process output is probably the priority measure of any process. Customer involvement in all phases of a process management program is positioned as critical.

Clearly defined process owners are also positioned as a critical dimension of the new model. Process ownership is discussed as an additional or alternative dimension of the formal organization structure. The difficulty in process ownership is that strategic business processes usually cut across boundaries of organizational power and authority as defined by the formal functional organization chart. Davenport suggested that during periods of radical process change, process ownership should be granted precedence. This will, in theory, grant the process owner legitimate power and authority across the interfunctional boundaries.

Davenport further defined a process perspective as a horizontal view of business that cuts across the organization with product inputs at the beginning and outputs and customers at the end. A process-oriented structure is defined as de-emphasizing the functional structure of business. The functional structure is positioned as having hand-offs between functions that are frequently uncoordinated. The functional structure also does not define complete responsibility and ownership of the entire process. No one is managing the ship, only pieces of it. This is expensive, time consuming, and does not serve customers well. The solution proposed is that the interfaces between functional or product units be improved or eliminated, and sequential flows across functions be made parallel through rapid and broad movement of information. Viewing the organization in terms of processes and adopting process innovation, as explained by Davenport, inevitably entails cross-functional and cross-organizational change. Just the identification and definition of these processes often leads to innovative ways of structuring work.

During the 1990s, many studies examined the issue of reengineering and business processes. The focus on business improvement in the 1990s was clearly on business process reengineering, re-orienting the organization toward processes, customers, and outcomes, as opposed to hierarchies. In most of the studies of technology-oriented reengineering, re-orienting of the people and the organization was the major challenge and opportunity for business improvement. Coombs and Hull reported in a 1996 research study an emergence of a "business process paradigm," a heterogeneous collection of theories, concepts, practices for analyzing organizations, and practices for managing organizations.[12] The authors suggested that, although these are as yet heterogeneous, they all share a common view of a fundamental change in managing and thinking about organizations. They are distinguished from previous forms of management and analysis in that the focus is no longer on optimizing the specialist functions within the organization (e.g., Operations, Marketing, HRM), but shifts instead to ways of understanding and managing the horizontal flows within and between organizations.

BUSINESS PROCESS ORIENTATION IN THE 90s: ORGANIZATIONAL DESIGN AND CULTURE

John A. Byrne, in the December 13, 1993 issue of *Business Week*, provided the popular foundation for this area of investigation when he described the old organizational model as a vertical organization,[13] an organization whose members look up to bosses instead of out to customers. Loyalty and commitment is given to functional fiefdoms, not the overall corporation and its goals. Too many layers of management still slow decision-making and lead to high coordination costs. The answer, said Byrne, is the *horizontal corporation*. The outcome of this model is said to be greater efficiency and productivity and is achieved by reengineering or process redesign. Byrne states that AT&T, Dupont, GE, Motorola, and many other firms are moving toward this model.

The horizontal corporation is described as eliminating both hierarchy and functional boundaries and is governed by a skeleton group of senior executives that includes finance and human resources. Everyone else is working together in multidisciplinary teams that perform core processes such as product development. It is suggested that an organization of this type would only have three or four layers of management between the chairman and the "staffers" in a given process. A stated goal of DuPont's is to get everyone focused on the business as a system in which the functions are seamless in order to eliminate the "disconnects and hand-offs." General Electric Chairman John Welch speaks of building a "boundary-less" company to reduce costs, shorten cycle time and increase responsiveness to customers. Managers in this organization would have "multiple competencies" rather than narrow specialties and would function in a group to allocate resources and ensure coordination of processes and programs. Byrne cited numerous examples of companies that are organizing around market-driven business processes and realizing cost reductions of 30% or more.

Byrne described the horizontal corporation model as a firm that has the following elements:

1. The company is built around three to five core processes, not tasks, with specific performance goals and a "process owner" assigned to each process.

2. The hierarchy is flattened. Supervision has been reduced, fragmented tasks combined, non-value-added work is eliminated, process activities are cut to a minimum, and as few teams as possible are used to perform an entire process.

3. Teams manage everything. Teams are the main building block of the organization with limited supervision by making the teams self-managed. The teams are given a common purpose and are held accountable for measurable performance goals.

4. Supplier and customer contacts are maximized. Employees are brought into direct, regular contact with suppliers and customers. In some cases, supplier or customer representatives are full-time working members of in-house teams.

5. All employees are informed and trained. Employees are trusted with raw data and trained how to perform analysis and make decisions.

6. Customers drive performance. Customer satisfaction, not stock appreciation or profitability, is the primary driver and measure of performance. The profits will come and the stock will rise if the customers are satisfied.

7. Team performance is rewarded. The appraisal and pay systems reward team results, not just individual performance. Employees are encouraged to develop multiple skills rather than specialized know-how and are rewarded for it.

With this article in 1993, Bryne popularized the term "horizontal organization" and provided a prescriptive definition of a business process-oriented model.

In an earlier work within the organizational design area, Rummler and Brache[14] proposed a framework based upon the premise that organizations behave as adaptive processing systems that convert various resource inputs into product and service outputs which it provides to receiving systems or markets. These organizations are based upon process-oriented structures, measures, rewards, and resource allocation.

Rummler and Brache suggested that the investments made in improving the firm using a functional orientation have resulted in functional optimization that suboptimizes the organization as a whole.[15] People in functional silos focus on what is best for that function, many times at the expense of other functions. This means that while the individual function benefits, oftentimes the firm as a whole loses. Figure 2.5 visually depicts their hypotheses of suboptimization.

To address the suboptimization phenomenon, Rummler and Brache suggest organizing jobs, structures, measures, and rewards around horizontal processes. This process-oriented organizational design is offered as

Cross-Functional Processes vs. Functional Silos

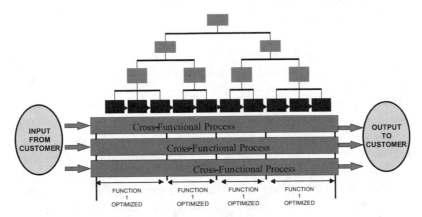

FUNCTION 1 OPTIMIZED   FUNCTION 1 OPTIMIZED   FUNCTION 1 OPTIMIZED   FUNCTION 1 OPTIMIZED

SUM OF OPTIMIZED FUNCTIONS ≠ OPTIMIZED PROCESS

Figure 2.5 The "Silo" Suboptimization Phenomenon

the improved model of business performance. In fact, during the 1990s, Rummler and Brache built a sizable consulting practice helping firms implement this model.

Along this same line, Melan from IBM published several articles in the quality literature suggesting the implementation of the principles of process management used successfully in manufacturing.[16] Melan suggested "viewing the operation as a set of interrelated work tasks with prescribed inputs and outputs" and provided a structure and framework for understanding the process and relationships and for applying the process-oriented tools used successfully in manufacturing.

Examples of these tools are the basic strategy of process measurement and control, statistical process control, cycle-time analysis and optimization, line balancing, variability analysis, reduction, and continuous process improvement. These strategies, tools, and techniques can only be successfully applied once a process-oriented framework is constructed.

Melan describes the application of these tools to a business process as process management. According to Melan, process management means establishing control points, performing measurements of appropriate parameters that describe the process, and taking corrective action on process deviations. Melan defines the six basic features of process management as:

1. Establish ownership of the process
2. Establish workflow boundaries
3. Define the process
4. Establish control points
5. Implement measurements
6. Take corrective action

Melan also strongly stated that the implementation of process management has the potential to yield operational improvements and should not be underestimated.

In 1997, researchers Detoro and McCabe defined business process management as the organizational improvement approach of the 1990s.[17] The current or functional view, as defined by Detoro and McCabe, is that a traditional organization is managed hierarchically; there is a chain of command where information flows upward to senior functional managers who evaluate the data, make decisions, and deploy policy and communications downward. Cross-functional issues are rarely addressed effectively and, consequently, the performance of the organization is suboptimized.

Future organizations, they said, will rely more heavily on horizontal, or business process management. In horizontal management, the organization is viewed as a series of functional processes linked across the organization, which is how work actually gets done. Policy and direction are still set at the top, but the authority to examine, challenge, and change work methods is delegated to cross-functional work teams. This "re-viewing" is the process of re-orienting the organization toward business processes.

Detoro and McCabe suggested that business process management solves many of the suboptimization problems in traditional structures because it focuses on the customer, manages hand-offs between functions, and avoids turf mentality because employees have a stake in the final result and not just what happens in their departments.

BPO, as defined by Detoro and McCabe, appears to be the restructuring and reviewing of the organization toward process, teams, and outcomes.

BUSINESS PROCESS ORIENTATION IN 2000: THE E-CORPORATION

The completion of the interstate highway system in the United States ushered in the age of transportation and made every business a national business. The completion of a usable global information network, the Internet, has made every company and market a global one, every customer an informed consumer, and brought us into a new economy, the "digital economy," with new rules and new realities.

The Internet has the capacity to change everything and is doing so at a far greater speed than the other "disruptive" technologies of the 20th century, such as electricity, the telephone, and the automobile. "In five years time, all companies will be Internet companies or they won't be companies at all," says Andy Grove, chairman of Intel.[18]

What is causing this major change in the way the world works? The list is long and somewhat speculative at this point, but some factors are becoming clear. The new assets are not factories, machinery, or raw materials but information, knowledge, relationships, and connectivity. Location, or "place" in the 4P marketing language, is becoming almost irrelevant and might be replaced with "perfection." "How you gather, manage and use information will determine whether you win or lose," says Bill Gates of Microsoft.[19]

Having information available to every customer, when and where they want it, at a cost affordable by almost everyone has dramatically shifted the balance of power and customer expectations. Customers, both end consumers and intermediaries, are expecting dramatically more: more information, more speed, more flexibility, more cooperation/collaboration, and more service. They are also expecting less: lower cost, less paper, fewer mistakes, fewer hassles. In the digital economy, they have the power to demand it all. Meeting these expectations and demands places a tremendous strain on our systems, people, organizations, and processes and has fundamentally changed the balance of supply and demand. Customers are in charge and information is the power. Understanding and leveraging this is the imperative for survival in the digital economy.

In a presentation to Wall Street analysts, Lou Gerstner of IBM described the new "dot-com" companies as "fireflies before the storm—all stirred up, throwing off sparks." But he continued, "The storm that's arriving—the real disturbance in the force—is when the thousands and thousands of institutions that exist today seize the power of this global computing and communications infrastructure and use it to transform themselves. That's the real revolution."[20] This means building the e-corporation.

What does this mean for business process orientation? As the e-forces compel the corporation to perform at even greater levels and focus outward on the customer, there can be no effort that is not value-added. With effortless globalization enabled by the Internet, competition increases exponentially. There can be no such thing as internally focused people and functional processes that bring little or no value to the customer. The only way to compete in this e-world is to become horizontal or business process oriented.

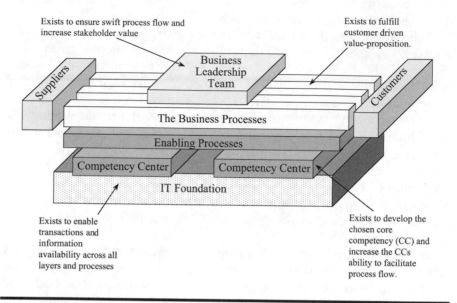

Exists to ensure swift process flow and increase stakeholder value

Exists to fulfill customer driven value-proposition.

Suppliers

Business Leadership Team

Customers

The Business Processes

Enabling Processes

Competency Center

Competency Center

IT Foundation

Exists to enable transactions and information availability across all layers and processes

Exists to develop the chosen core competency (CC) and increase the CCs ability to facilitate process flow.

Figure 2.6 The BPO E-Corporation

For example, hundreds of companies are now forming that exist solely around a business process: e-procurement. This totally business process–oriented organization can operate at efficiencies that are 10 to 20 times those of the functional, internally focused model. These are only the first of the many BPO e-corporations yet to come.

What do these e-corporations look like? Figure 2.6 offers one possible view.

This totally horizontal view ignores traditional ownership boundaries and geographies. This view could include hundreds of legal entities and span the globe. The functions only exist as competency centers and these could also be different legal entities. The leadership is in the form of a team representing the stakeholders: the legal shareholders as well as customers, suppliers, and participants in the e-corporation.

It is apparent from this brief description and view of the e-corporation that BPO is the fundamental orientation guiding the building and operation. Therefore, defining, measuring, and exploring the impacts of BPO become even more important today.

SUMMARY

The BPO commonalties in the literature appear to be centered on a "process culture" with structures and systems consistent with that culture.

A "systems" approach is also clearly a common component of BPO as is the integration of the entities outside of the formal organization (suppliers and customers). The literature stresses that customer focus is a strong part of this "process culture."

A "business process culture" is a culture that is cross-functional, customer oriented along with process and system thinking. This can be expanded by Davenport's definition of process orientation as consisting of elements of structure, focus, measurement, ownership, and customers. Commitment to process improvement directly benefits the customer and business process information-oriented systems as a major component of this culture.

A culture of "teaming" and empowerment is critical for the practice of a business process orientation. This teaming culture consists of empowered individuals focused on customer value and continuous improvement of both results and processes. Integrating mechanisms such as teaming, reward systems, and information are also key elements driving business process orientation.

Finally, both cross-functional and outcome-oriented process thinking is needed. A process-driven organization can be characterized by such major components as business processes, jobs and structures, management and measurement systems, and values and beliefs.

NOTES

1 Payne, A. F. Developing a marketing-oriented organization, *Bus. Horizons*, May–June, 1988, pp. 46–53.
2 Porter, M. E. *Competitive Advantage: Creating and Sustaining Superior Performance*, New York, The Free Press, 1985.
3 Walton, M. *The Deming Management Method*, New York, Perigee Books, 1986.
4 Davenport, T. H and Short, J. E. The new industrial engineering: Information technology and business process redesign, *Sloan Manage. Rev.*, 31, 1990, pp. 11–27.
5 Hammer, M. and Champy, J. *Reengineering the Corporation: A Manifesto for Business Revolution*, New York, HarperBusiness, 1993.
6 Drucker, P. F. *The New Realities*, New York, Harper & Row, 1989.
7 Walton, M. *The Deming Management Method*, New York, Perigee Books, 1986.
8 Imai, M. Kaizen: *The Key to Japan's Competitive Success*, New York, McGraw–Hill, 1986.
9 Drucker, P. F. The coming of the new organization, *Harvard Bus. Rev.*, Jan.–Feb., 1988, pp. 45-53.
10 Hammer, M. Reengineering work: Don't automate, obliterate, *Harvard Bus. Rev.*, July–August 1990, pp. 104–112.
11 Davenport, T. H. *Process Innovation: Reengineering Work through Information Technology*, Boston, Harvard Business School Press, 1993.
12 Coombs, R. and Hull, R. The wider research context of business process analysis. (Working Paper) Center for Research on Organizations, Management and Technical Change, Manchester School of Management, Manchester, U.K., 1996.

[13] Byrne, J. A. The horizontal corporation, *Bus. Week,* December 13, 1993, pp. 76–81.

[14] Brache, A. P. and Rummler, G. A. *Improving Performance: How to Manage the White Space on the Organizational Chart,* San Francisco, CA, Jossey-Bass, 1990.

[15] Byrne, J. A. The horizontal corporation, *Bus. Week,* December 13, 1993, pp. 76–81.

[16] Melan, E. H. Process management in service and administrative operations, *Qual. Prog.,* 1985, pp. 52–59.

[17] Detoro, I. and McCabe, T. How to stay flexible and elude fads. *Qual. Prog.,* Vol. 30, March 1997, pp. 55–60.

[18] Staff, The net imperative, *The Economist,* June 26, 1999.

[19] Gates, B. *Business @ the Speed of Thought,* New York, Warner Books, 1999.

[20] Staff, The real revolution, *The Economist,* June 26, 1999.

3

DEFINING AND MEASURING BPO

As described in the previous chapter, the concept of business process orientation (BPO) has only been generally defined and not measured or tested to determine its impact on an organization. We also concluded that there appears to be a general consensus as to the components of BPO. Yet, to date, no one has developed and tested this concept. In order to take the understanding and implementation of BPO further, the development of a concise definition of BPO and a qualitative measurement instrument was needed.

Why are BPO definitions and measures needed? If you cannot clearly define, describe, and measure something, you will not know if you ever have it. If you cannot determine the impact of it, you may not even be sure you want it. In other words, you cannot manage what you cannot measure. With this in mind, a multi-year research study was begun in 1996 to develop and test a valid and reliable BPO measure. The steps involved and the results of this study are discussed below.

STUDY OVERVIEW

During the early 1990s, the BPO concept attracted significant practitioner and researcher interest, and was implemented in whole or in part by enough companies that a substantial experience pool became available. The challenge, of course, was tapping this broad range of experience and distilling it into a format that practitioners could use and easily understand.

The approach we took was to review the popular business press and to interview experienced practitioners and experts both in the United States and Europe to help define BPO and its major components. Various statistical techniques (domain sampling, coefficient alpha analysis, and factor analysis) were used to produce a more parsimonious measure of BPO and to elicit its major dimensions.[1]

Key informant research was used to investigate the process orientation of selected organizations in the United States during 1998. A key informant study is one that selects participants based on their level of understanding about a certain topic. In this case, early on in the study, participants were selected on the basis of their understanding of the BPO topic and during the later phase of the study selection was based upon BPO understanding and their understanding of the organization to which they belonged.

Our research was divided into two phases. Phase 1 involved developing a valid measure of BPO. Phase 2 involved testing these measures using a large national sample by administering a self-assessment questionnaire to gather data from several judgmental samples (population selected based upon certain criteria).

Phase 1: Developing BPO Definitions and Measures

The objective of Phase 1 was to generate a validated definition of BPO and produce a valid and reliable measurement tool to be used in future research. This effort began with the development of several definitions derived from an extensive review of the published literature on the subject. A list of 200 possible measures was developed during this review. A Delphi technique was used by sending the preliminary questionnaire to several BPO experts around the world for a "reality" check. Feedback from this jury of experts was used to prune and refine the BPO measure.

In keeping with the Delphi approach, a list of questionnaire items recommended by the jury of experts was then distributed to several hundred practitioners for their review. The participants were asked to numerically rate each definition and candidate measure according to its relevance in defining BPO. A 5-point Likert scale was used with 1 indicating completely disagree with the relevance of the questions or measure and 5 indicating completely agree. The results were then examined. The definitions and questions with the average relevance scores below a rating of 3 were removed from the list.

Once the data were collected and analyzed, a consensus of two definitions of BPO seemed to surface:

> An organization that is oriented toward processes, outcomes, and customers as opposed to hierarchies.
> An organization that emphasizes process and a process-oriented way of thinking.

These two definitions were then combined to most accurately represent the BPO construct. Thus, the final definition of BPO to be used in all future research can be stated as follows:

An organization that, in all its thinking, emphasizes process as opposed to hierarchies with special emphasis on outcomes and customer satisfaction.

The list of measures resulting from this Delphi process contained 200 questions organized into five major categories, representing dimensions of a BPO organization including:

1. A *process view* of the business,
2. *Structures* that match these processes,
3. *Jobs* that operate these processes,
4. *Management and measurement systems* that direct and assess these processes, and
5. Customer-focused, empowerment- and continuous-improvement-oriented *values and beliefs* (culture).

This new list of questions or possible measures of BPO was again distributed to several hundred practitioners. The participants were asked to rate these items again using a 5-point Likert scale measuring agreement with the question in regard to the participant's organization. This scale consists of the following response categories:

1 — completely disagree
2 — mostly disagree
3 — neither agree nor disagree
4 — mostly agree
5 — completely agree
8 — cannot judge

The data were collected and analyzed. Statistical data reduction techniques produced a much more "elegant" BPO measure, consisting of three broad dimensions and 11 survey questions out of the original 200 items. The three dimensions that composed the final survey instrument were: **Process jobs** (PJ), **Process management and measurement** (PM), and **Process view** (PV). To ensure the confidence in utilizing the BPO scale to accurately assess an organization's process orientation, a second phase involved refining the scale's properties. (See Appendix C for the final version of the BPO questionnaire.)

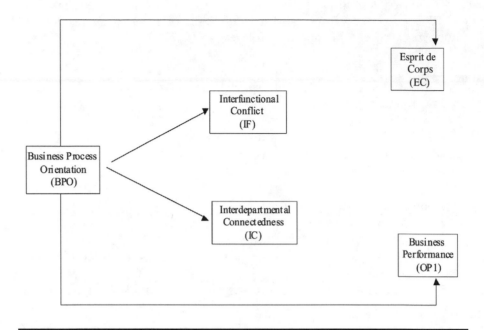

Figure 3.1 BPO and Organizational Impact Variables

Phase 2: Expanded Testing and Further Validation

The objective of Phase 2 of this study was to further test and validate the definitions and measurements. This was accomplished in two ways. First, identical tests from Phase 1 were performed on a larger sample to see if the results matched. Second, BPO was compared to several organizational variables to determine if the proposed relationship of BPO to these organizational variables actually matches statistical reality. The organizational impact variables examined were the conflict and connectedness between functions within and organization, the overall business performance of the organization, and the overall feelings of esprit de corps in the organization. All of these organizational variables were thought to be logically linked to BPO. The logic of the BPO measures can be tested by examining these relationships utilizing a "Does this make sense?" test. If BPO can be shown to increase or decrease in a way that tracks the increase or decrease of other organizational factors that have already been defined and tested, then the BPO measures can be said to have a logical or face validity.

The relationship to these organizational factors will be examined in greater detail in the following chapters. Figure 3.1 shows the proposed variables that relate to BPO in an organization.

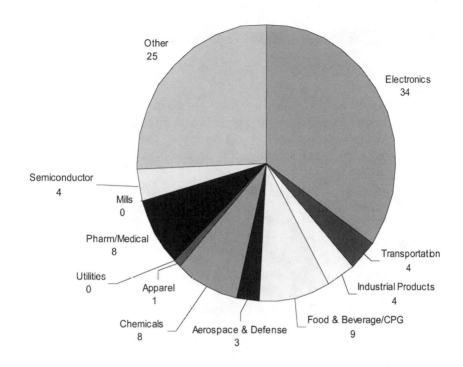

Figure 3.2 Industries Represented in the Sample

As a proposed "Does this make sense?" logic, it was thought that BPO would reduce conflict between functions, improve the connectedness between these functions, improve overall business performance and increase feelings of esprit de corps. All of these factors had well-established measures that could be included in the BPO survey instrument and, thus, were easily distributed with the BPO questions.[2]

For the final data gathering in Phase 2, a judgmental sample of participants was selected from Hammer and Co. reengineering seminar attendee lists based upon company type; manufacturing firms in the U.S. were the unit of analysis. Data were gathered from participants at a cross-company internal Motorola seminar.

Approximately 500 survey questionnaires were distributed by regular and electronic mail to the list of participants. A total of 115 responses were received and subject to identical statistical tests used in Phase 1. Figure 3.2 shows the industries responding to the survey as very broad, with a strong concentration in the electronics.

Furthermore, in order to ensure that one department or function was not overly represented, a functional distribution of the sample respondents was conducted. As Figure 3.3 shows, no single function was overly represented.

Respondents by Function

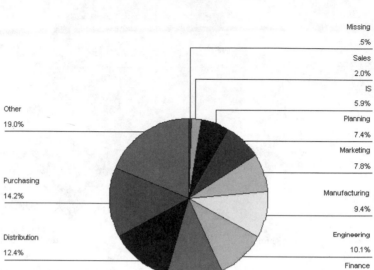

Figure 3.3 Functional Breakdown of Survey Respondents

A final check was conducted on whether the survey respondents were fairly represented based on the management levels of the survey respondents. The data were examined for the number of respondents by position in their organization, as shown in Figure 3.4. The respondents are broadly distributed across different levels, from individual contributor to senior leadership. Based on the foregoing analyses of the sample respondents, we concluded that the data collected were generally representative of the manufacturing organizations in the United States and would provide conclusions that would be broadly applicable to this population as a whole.

These data became the basis for examining BPO and its impact on the organization, which will be discussed further in the following chapter.

Using statistical data reduction techniques, the statistical relationships from Phase 2 were compared to the results in Phase 1. The results in Phase 2 duplicated the Phase 1 results and thus provided the first level of validation for the measurements. The relationship statistics of the

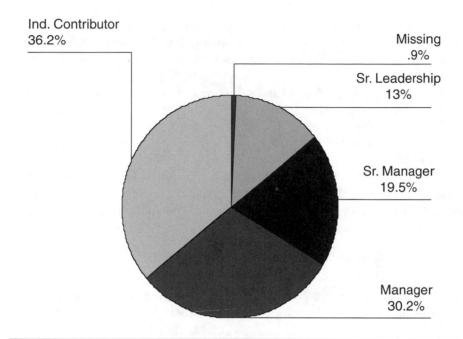

Ind. Contributor
36.2%

Missing
.9%

Sr. Leadership
13%

Sr. Manager
19.5%

Manager
30.2%

Figure 3.4 Positions Responding to the Survey

measures to each other and to overall BPO were almost identical thus providing validation that the measures were operating in a repeatable way.

For the second level of validation, whether the proposed relationships match, statistical tests were used to test the proposed relationships between BPO and the impact factors shown in Figure 3.1. The relationships between BPO and esprit de corps and business performance were all strong, significant, and in the right direction, meaning that when BPO increases, business performance and esprit de corps increase. The relationships to interdepartmental (or interfunctional) conflict and connectedness were also strong and significant. Interdepartmental conflict was considered to have an inverse or negative relationship with BPO, and this was shown to be true.

The conclusions from Phase 2 were that the measurement or survey instrument was actually measuring BPO and could provide repeatable results. Additional validation of the measures and the BPO concept was provided by the fact that the organizational relationships logically proposed as a "Does this make sense?" validation test came through in the statistical testing. Actual use of the BPO definition and measures seem, at least on the surface, to make sense.

CONCLUSION

A BPO is unequivocally crucial to long-term business success, yet to date, no valid measure exists to assess its properties. Our research offers not only a robust definition of BPO, but also a valid measure of this important construct.

This measure has been developed using experts from around the world and a significantly representative test, both from an industry functional and position perspective. In this test, the measures have also passed the "Does this make sense?" test of a logical relationship to other organizational variables.

The creation of this measurement mechanism will enable the deeper examination of BPO as a practical concept that could result in significant organizational performance improvements. The following questions can now be fully examined:

> What is BPO?
> How do I know when I have it?
> What are the impacts of BPO on my organization?
> Can BPO make a competitive difference?

Using the data set gathered in Phase 2 of this research and discussed in this chapter, Chapter 4 will examine the impact of BPO on organizational performance.

NOTES

[1] McCormack, Kevin (1999). The development of a measure of business process orientation and its link to the interdepartmental dynamics construct of market orientation, *Diss. Abstr. Int.,* DAI-A 60/07, January 2000, p. 2589.
[2] Jaworski, B. J. and Kohli, A. K. Market orientation: Antecedents and consequences, *J. of Mark.*, Vol. 57, 1993, pp. 53–70.

4

BPO AND ORGANIZATIONAL PERFORMANCE

We argued at the outset that a BPO represents a powerful force in transforming the organization. However, apart from hard data, we have been unable to state conclusively that BPO really makes a difference. Chapter 3 described the development of BPO as a measurable concept. This chapter reports on actual research conducted using the BPO survey instrument in a cross-industry, key informant study to determine how BPO affects an organization.

DOES BPO MATTER?

Figure 4.1 shows four potential outcome variables selected to answer the question whether BPO is related to improved organizational performance and long-term health. These are overall business performance, interfunctional conflict, interdepartmental connectedness, and esprit de corps. These factors were selected based upon their use in previous research, where they had also been significantly defined and measured.[1]

The proposed internal organizational impacts of BPO are interfunctional conflict and interdepartmental connectedness. Interfunctional *conflict* is defined as tension among departments arising from the incompatibility of actual or desired responses and interdepartmental *connectedness* as the degree of formal and informal direct contact among employees across departments. An increase in conflict across functions is thought to be a negative internal organization factor. Incompatible goals and tension between individuals in different functions, sales and manufacturing, for example, have been shown to negatively impact

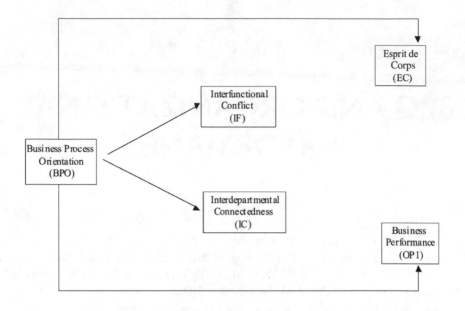

Figure 4.1 Proposed Organizational Impact of BPO

organizational performance. An increase in connectedness across departments as measured by the easy flow of communication between departments and a low level of tension among members of each department has been shown to contribute to improved organizational performance.[2]

Implementing BPO as a way of organizing and operating in an organization will improve internal coordination and break down the functional silos that exist in most companies. Research has shown that this increase in cooperation and decrease in conflict improve both short- and long-term performance of an organization.

Overall Business Performance

Organizational performance can vary greatly among companies competing in similar markets. Moreover, industries apply different performance metrics, making cross-industry comparisons difficult. For example, the retail industry uses rapid inventory turns as a key performance metric in measuring good performance, while the defense industry defines good performance as something very different. For this reason, we selected a self-report rating system to measure overall performance of the organizations studied. Use of key informant self-ratings has been shown to be closely approximate quantitative measures of performance and can also be used to compare organizations in different industries. Research has also shown that key informants can accurately and honestly position their organizations on an objective performance scale.[3] Using a 5-point rating scale, participants in our research were asked to rate their own organizations' performances as well as those of their competitors.

The overall long-term health of an organization can be predicted from the attitude of the members. Team spirit and feelings of "being in it together," generally described as *esprit de corps*, have been shown to be the energy and glue of an organization. Esprit de corps is defined as a spirit of enthusiasm and devotion to a common cause among group members. This team spirit is the subject of thousands of leadership books, tapes, and speeches. Unfortunately, the restructuring and downsizing of the 1980s and 1990s destroyed this spirit and organizations have spent many millions of dollars to attempt to rebuild this team spirit. Many leadership heroes and gurus have made their reputations by building this spirit of enthusiasm and credit their success as leaders to this ability. Consider, for example, Southwest Airlines, the number one airline in almost every performance and customer satisfaction measure. A strong esprit de corps instilled by its charismatic leader, Herb Kelleher, has made Southwest profitable for 26 straight years, with an average EBITDA margin of 22.6%.[4]

RESEARCH FINDINGS

The data gathered in Phase 2 of the measurement development and validation project was used as the research sample. This was a judgmental sample of participants selected from Hammer and Co. reengineering seminar attendee lists based upon company type, with manufacturing firms in the United States as the unit of analysis. This database consisted of 115 responses from a broad cross-section of industries, functions, and positions within organizations (from CEO to individual contributor). The respondent companies also varied in size from approximately $100 million to several billion in annual sales.

We used a three-step process to gauge the affect of BPO on organizational climate and performance. First, we prepared a simple correlation matrix to determine the strength of association between BPO and organizational climate and performance. We found that BPO and esprit de corps (EC) had a strong, positive correlation, indicating that BPO can dramatically influence the health of an organization as described by the employees' feelings of enthusiasm and devotion to a common cause.

We also found that BPO and interdepartmental connectedness showed a fairly strong positive correlation. This indicates that the cooperation across departments increases as BPO increases in an organization. On the other hand, BPO and interfunctional conflict exhibited a strong inverse relationship, indicating that when BPO increases, conflict across functions decreases.

These are important findings. Many companies have spent millions of dollars on cross-functional teaming programs and consultants with the intent to improve interdepartmental cooperation and reduce conflict. Most companies have done this without changing their organizational structures. They have maintained the functional, departmental silos and attempted to overcome the conflicts by asking everyone to work on cross-functional teams and "all get along." In some companies, a single person could be on 20 teams, an impossible situation with questionable results.

Finally, we wanted to see what effect BPO would have on overall business performance. Our results indicate a surprisingly strong relationship between BPO and overall performance. Considering all the factors that can potentially affect business performance, this finding is compelling. Business schools at most colleges and universities today focus on strategies and tactics that lead to successful business performance. How can one factor, the BPO of an organization, have this much impact on overall business performance?

When analyzing our data according to each dimension (Process Jobs, Process View, Process Management and Measures) of BPO, the results yielded several interesting findings. First, the general relationship between all the components and interfunctional conflict is negative. This shows that as the BPO components of Process Management and Measures (PM), Process Jobs (PJ), and Process View (PV) increase, interfunctional conflict should decrease. Second, the relationship between PM and all the impact factors is stronger than the other BPO components. This indicates that PM may be the most important component of BPO. Process-oriented measures are, by definition, cross-functional and logically, this would contribute toward a common cause required for EC to occur. Finally, PJ seems to be the next important component with PV the least important but still significant (see Table 4.1).

Table 4.1 Correlation Matrix: BPO Dimensions and Outcomes

Impact Factors	Component 1: PM	Component 2: PJ	Component 3: PV
Conflict–IF	–0.325*	–0.231*	–0.279**
Connectedness–IC	0.309*	0.262**	0.187**
Performance–OP1	0.319*	0.206*	0.111***
Esprit de Corps–EC	0.428*	0.313*	0.308*

*Significant at the 0.01 level. **Significant at the 0.05 level. ***Significant at P = 0.248.

Is there a logical explanation for this ranking? For PM, it might be said that what gets measured and rewarded gets done. Having horizontal or process measures that require groups of people to work together toward common goals should build a team spirit. Assuming the right measures were being used, this would positively affect the bottom line. Given the strong relationship between PM and overall performance (OP1), this seems to indicate that process-oriented measures contribute to greater overall performance. The strong relationship to EC and the conflict and connectedness factors indicate that process-oriented measures will decrease conflict, increase connectedness, and result in an increase in the overall feeling of EC. Thus, PM seems to exert disproportionate influence, both on building a strong organizational culture (i.e., lower conflict, stronger sense of connectedness) and improving overall company performance.

Process jobs (PJ), the BPO factor exhibiting the next greatest influence, also seems to make sense. With "jobs" comes authority. With process-oriented jobs comes "horizontal" or process-oriented authority. This type of authority would span functional boundaries and, by definition of BPO, use of this authority would encourage employees from different functions to work together toward common goals. Employees reporting to a common manager are very likely to work toward common goals. Therefore, creating strong relationships that reduce conflict, increase connectedness, and build EC all seem to make sense. The relationship to overall performance also makes sense. When there is unity of purpose around a leader, this energy usually translates to an improved bottom line.

The relationship of process view (PV) to the other organizational factors is the weaker when compared to the other BPO dimensions. Quite simply, PV is about documentation and understanding. Using a cross-functional team to collectively view and describe process activities and responsibilities promotes EC by virtue of working side by side. It is a team-building exercise. This activity and the clarity of roles can also reduce conflict and improve cooperation. Should this working together lead to new or redesigned processes, the result will be significantly higher connectedness within the organization.

The relationship of PV to overall performance seems a little weak. A possible explanation is that creating documentation and understanding by itself will not lead to higher overall business performance. Only the commitments among team members created during this documentation process can lead to performance improvements. This process includes not only documenting a process but a team agreeing upon what activities are performed, how they will be measured, and who is responsible for the process outcomes. This system of agreements is the foundation for everything else that needs to be built to become business process oriented. It is always difficult to measure the direct contribution of a foundation, by itself, to performance.

DISCOVERING THE MAGNITUDE OF THE RELATIONSHIPS

How does this all fit together? To answer this, we applied regression analysis to our data to examine the strength of relationship of individual factors affected by a BPO.

Our propositions were that BPO helps improve overall business performance (OP1), reduces interfunctional conflict (IF), and improves interdepartmental connectedness (IC) and esprit de corps (EC). The overall model and regression coefficients for each relationship in the model are shown in Figure 4.2.

These results are extremely encouraging for several reasons. First, our proposed BPO-performance link was quite strong (0.279, p<0.003). We now have empirical data that support the proposition that BPO leads to greater overall business performance. Second, we were quite pleased with how the research supported the "directionality" of our model. That is, the behavior of the factors and their interrelationships in our model performed as expected. The logic of how BPO drives organizational culture (e.g., reduced interfunctional conflict and greater interfunctional coordination) was clearly shown. Managers can use this model as a prototype as they reconfigure or reinvent their organizations to effectively compete in the new millennium.

CONCLUSIONS

This research has captured the impacts of BPO on an organization. BPO can contribute to the overall performance of an organization by reducing conflict and improving interdepartmental cooperation. BPO has also been shown to improve EC in an organization. Companies structured into broad process teams rather than narrow functional departments should have less internal conflict and stronger team spirit. Furthermore, BPO is not dependent on a charismatic leader but on the fundamental organizational dynamics. It will not change when the leader leaves, as often happens. This provides a sustainable approach to improved performance and organizational health.

Process measurement (PM) and process jobs (PJ) had strong relationships to all the organizational variables we studied, while process view (PV) did not. We interpret this to mean that documentation alone (i.e., PV) does not have a major impact. Documentation merely provides a foundation that can be used to organize jobs and measures. For example, TQM efforts during the 1980s stressed thorough process documentation,

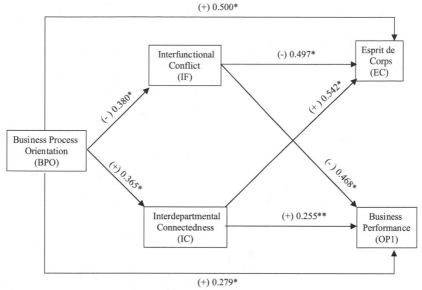

Note: All numbers shown are Standardized Regression Coefficients of relationships.
Significance: ** p<.007, * p<.001

Figure 4.2 Regression Analysis Results

assuming that this was key to organizational success and that process view alone might carry the day, but it did not.

Chapter 5 will develop the concept of BPO maturity and introduce a maturity model that helps evaluate the level of BPO in an organization. This model also helps highlight focus areas of effort needed for an organization to move forward on this journey toward a business process-oriented organization.

NOTES

[1] Jaworski, B. J. and Kohli, A. K. Market orientation: Antecedents and consequences, *J. of Mark.*, Vol. 57, 1993, pp. 53–70.

[2] Jaworski, B. J. and Kohli, A. K. Market orientation: Antecedents and consequences, *J. of Mark.*, Vol. 57, 1993, p. 57.

[3] Rodgers, E.W. and Wright, P. M. Measuring organizational performance in strategic human resource management: problems, prospects and performance information markets, *Hum. Resour. Manage. Rev.*, Fall 1998, pp. 314–320.

[4] How Herb keeps Southwest hopping, *Money*, June 1999, p. 61.

5

BENCHMARKING USING THE BPO MATURITY MODEL

Chapter 4 discussed the results of several years of research on the impact of BPO on an organization. Clearly, becoming business process oriented in an organization represents a significant challenge yet confirms potentially high returns. Defining the end goal of this journey and finding out where you are on this journey is critical. This chapter answers these questions by describing how to use the BPO measurement instrument to position an organization on a BPO maturity model.

WHAT IS BPO MATURITY?

In creating a BPO maturity model we defined maturity as the stages through which an organization progresses in becoming business process oriented, ultimately realizing an end goal of being fully process integrated. A major inspiration for the model comes from Philip Crosby, who developed a maturity grid for the five stages that companies go through in adopting quality practices. Crosby suggested that small, evolutionary steps, rather than revolutionary ones, are the basis for continuous process improvement. We believe the same holds true for BPO. Each successive step includes more practices involving more functions and more people within a given organization.

As we analyzed the data collected from 1997 through 1999, we saw patterns and clear evolutionary stages. After examining these patterns and the stages as quantified by the BPO measurement instrument, we developed the following definitions and numerical ratings (0 to 5) for the stages that an organization goes through to become business process oriented.

■ *Ad Hoc:* The processes are unstructured and ill defined. Process measures are not in place and the jobs and organizational structures are based upon the traditional functions, not horizontal processes. Individual heroics and "working around the system" are what make things happen across functions and departments. This is BPO stage as defined by a BPO score of 0 to 2.

■ *Defined:* The basic processes are defined, documented, and available in flow charts. Changes to these processes must now go through a formal procedure. Jobs and organizational structures include a process aspect, but remain basically functional. Functional representatives (from sales, manufacturing, etc.) meet regularly to coordinate with each other, but only as representatives of their traditional functions. Similarly, functional representatives meet to coordinate activities with vendors and customers. Companies reaching this level of process maturity record scores of 2 to 3.

■ *Linked:* The breakthrough level. Managers employ process management with strategic intent and results. Broad process jobs and structures are put in place outside of traditional functions. One common indicator is the appearance of the title "process owner." Cooperation between intracompany functions, vendors and customers, takes the form of teams that share common process measures and goals that reach horizontally across the company. A BPO score of 3 to 4 characterizes firms at this level of BPO maturity.

■ *Integrated:* The company and its vendors and suppliers take cooperation to the process level. Organizational structures and jobs are based on processes, and traditional functions begin to be equal or sometimes subordinate to process. Process measures and management systems are deeply imbedded in the organization. Firms that score 4 to 5 we define as truly integrated, having achieved optimal balance between process and function.

Figure 5.1 shows the four levels of BPO maturity with the horizontal aspects, or process orientation, becoming more clear and powerful and the vertical functional aspects becoming less dominant as an organization progresses along the maturity path. At the highest level of maturity, the horizontal appears as strong an orientation as the functional or vertical orientation. In the maturity model, this is shown as strong, heavy horizontal lines crossing strong, heavy vertical lines. This represents the ultimate balance between process and functional orientation.

Here is how to interpret the maturity model shown in Figure 5.1. At the Ad Hoc level, the horizontal processes are barely visible and the functions or vertical silos are strong and clearly the way the organization is viewed, structured, and measured.

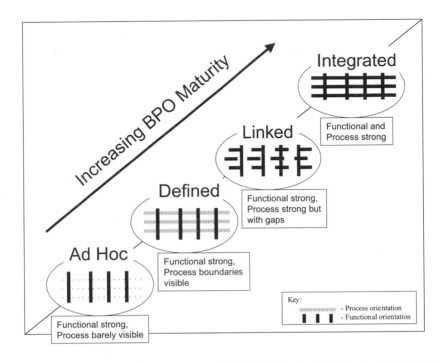

Figure 5.1 The BPO Maturity Model and the Phases of BPO

The Defined level is shown as still having dominant functions but the horizontal processes are slightly more visible with a process structure and measures beginning to appear. The Linked level clearly shows horizontal processes defined and structured almost on par with the vertical functions but there are gaps in the processes. At the Integrated level, the horizontal processes are strongly visible and on par with the vertical functions. There are no gaps, the horizontal structure is clear, and process measures help provide a horizontal focus.

By combining the BPO measurement tool and the maturity model, we can graphical place an organization on this continuum. Since the measurement tool produces an aggregate rating for an organization from 1 to 5, we can calculate the individual rating of an organization from the surveys completed within that organization. Using the results of our earlier research, we developed numerical scores that seem to correspond to the different maturity levels as described earlier.

Figure 5.2 shows the midpoint scores for each level of BPO maturity. A score of 0 to 2 would put and organization at the Ad Hoc level, of 2 to 3 at the Defined level, of 3 to 4 at the Linked level, and of 4 to 5 at the Integrated level.

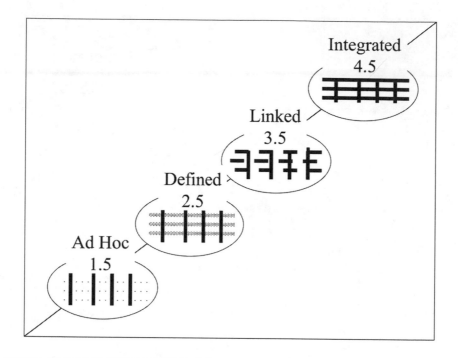

Figure 5.2 BPO Maturity Scores by Level

In this way, the views of the key informants in an organization can be gathered. Using the tool and aggregate scores can provide the current location of an organization on this "Mall Map." Like the locators at the entrance of most shopping malls, an organization can plot a "You are here" point and understand how far it has come and plot its journey forward.

BENCHMARKING USING THE BPO MATURITY MODEL

An organization that is trying to become business process oriented can sometimes find it helpful to compare itself to other organizations. The BPO measurement tool and the maturity model can be very useful in this effort and the current database of firms that have used the tool to measure their progress is close to 100 and growing.

Three steps are involved in this benchmarking effort.

1. Gathering the initial data and plotting the results on the high-level maturity model.

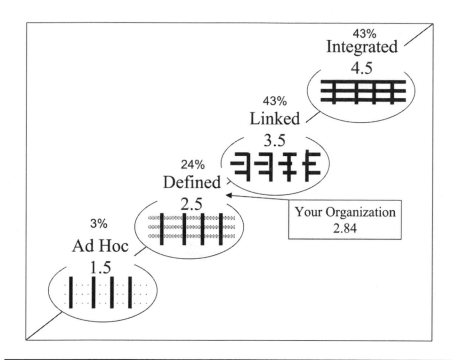

Figure 5.3 Benchmarking an Organization's BPO

2. Examining individual BPO component/outcome scores using BPO maturity model.
3. Comparing the detailed answers to the benchmarking database.

The first step in conducting this comparison or BPO benchmarking is to gather the data on your specific organization. This is accomplished by selecting 20 to 30 "key informants" within an organization that can complete the BPO measurement survey (see Appendix C). After aggregating and averaging the answers, the benchmarking begins by plotting the BPO score on the high-level maturity model show in Figure 5.3. For example, if the overall BPO score were 2.84 then the organization would be positioned as shown, slightly beyond the Defined stage. The percentages shown on the high-level maturity model represent the percentages of respondents at each stage in the database.

The next level of comparison could be undertaken by looking at the score for each component of BPO and its outcome. This would include the BPO components of Process View (PV), Process Jobs (PJ), and Process Management and Measures (PM) and the impacts of Interfunctional Con-

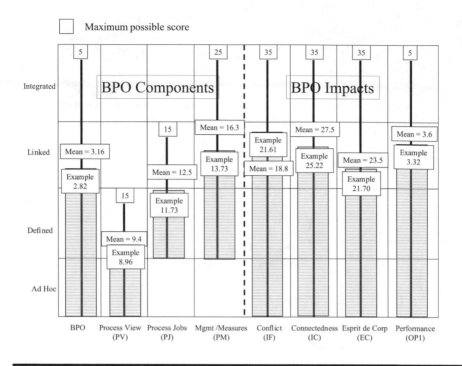

☐ Maximum possible score

Figure 5.4 Detailed BPO Maturity Model Example

flict (IF), Interdepartmental Connectedness (IC), Esprit de Corps (EC), and Overall Business Performance. Figure 5.4 is a more detailed maturity model used for this comparison.

The detailed model has the total possible and database mean score for each component and impact. The model also indicates the contribution of a BPO component to each level on the maturity scale. For example, PV can move an organization from Ad Hoc to Defined but, by itself, no further. Therefore, the scale for PV on the model ends at the top of Defined. Another example is PJ and PM. This effort toward BPO cannot even begin until an organization has the first portion of PV under way since the processes have to be first identified at a basic level before jobs and measurements can be put in place. For this reason the scale for PJ and PM does not even begin until the Defined stage is reached.

Figure 5.4 contains an example of how the second-level benchmarking results would be placed on the detail maturity model. The vertical bars represent the average scores of the organization being examined. The left side of the model contains the BPO components and the right side contains the outcomes. A numerical score is shown at the top of each bar. This is the average score for this category. The total possible score is shown at the top of the scale and the mean is shown in the center of the scale.

In this example, this organization is slightly above the Defined stage, as pointed out with the high-level maturity model. The PM component is clearly leading the way while the PV component is lagging. This company has "put the cart before the horse." A low PV score might suggest that many people in the organization do not understand the processes in the organization or that not all the processes are documented to a level where people can understand them.

The impact section seems to be in line with an organization in the early stages of the Linked level with conflict (IF) being the most leading. This scale is reversed from the actual scores since IF has an inverse relationship with the other variables. As BPO increases conflict is reduced, but in this case, since we reversed the scales, high IF numbers are good.

The third level of benchmarking and analysis is accomplished by looking at the answers to specific questions within the BPO measurement tool. Since PV seems to be an issue with this example, the answers to the three questions around PV will be examined. Figure 5.5 shows the answers for this specific organization compared to the database. The horizontal bars indicate the mean of the answers in the database and the diamond positions indicate the mean of the specific organization being examined. The percentage of respondents answering each question with

Business Process Orientation Survey

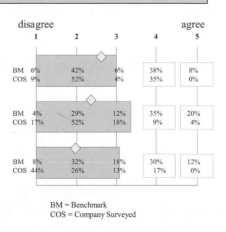

Figure 5.5 Benchmarking Detailed Questions regarding PV

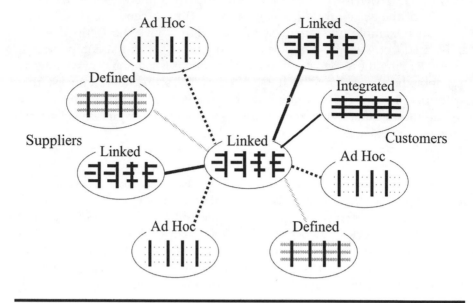

Figure 5.6 BPO Network Compatibility for the E-Corporation

a 1, 2, 3, 4, or 5 (where 1 indicates "completely disagree" and 5 indicates "completely agree") is also shown just above the bar.

The detailed answers to the PV questions show clearly that this organization is well under the mean in all three areas of employee process: orientation, the use of process terms, and organizational understanding of the processes. By sorting the individual answers by groups, functions, or titles within the organization the problem areas can be further identified and action can be taken to improve on this BPO component.

USING BPO MATURITY TO ALIGN THE E-CORPORATION

Competition in today's economy is no longer between individual companies but between groups of companies organized in value networks. According to Northwestern University Professor Philip Kotler, "Increasingly, competition will not be between companies but rather between marketing networks."[1] Partnerships, alliances, joint ventures, and cross-company collaboration are all basic modes of organizing business networks. With the dawn of the digital and Internet age, the next level of competition is cross-company process integration across this network.

A major challenge for competing in the Internet economy is to achieve BPO compatibility among the companies involved in the network. Figure 5.6 is a picture of a simple network complete with the level of BPO maturity of each participant. This example shows a company that is the hub of the network, which adds value by bringing together suppliers and customers. This could be a retail company, a distribution company, or a value-added bolt manufacturer. A likely scenario for today's e-corporations is that the network is made up of supplier and customer companies representing each of the BPO categories: Ad Hoc, Defined, Linked, and Integrated. The company in the center of this network, or the builder of the network, also has its own BPO level. In this case the company's BPO level is Linked.

How can all these different BPO "personalities" interoperate successfully at the process level? The answer is that each cross-company process in the network must be tuned to the lowest level of compatibility. In this example, the organizing company in the center is at the *Linked* level and one of its customers is at the Integrated level. The process connecting the two must be limited to the Linked level since the organizing company cannot operate at a level any higher than this. Process shock absorbers (usually people who span the two organizations) must be used to balance the interactions.

On the supplier side, the organizing company in the center must build links tuned to each supplier. An Ad Hoc supplier must be linked with an Ad Hoc compatible link and the Defined with a Defined link. If this is not in balance, the network will be ripped apart due to process incompatibility.

From this example, you can see how the BPO measurement system can be used as a critical strategy tool for building the e-corporation networks for today's economy. Assessing the BPO maturity level of all the partners and customers is a major prerequisite for building a successful network.

CONCLUSION

The BPO measurement tool and maturity model can be a useful tool to determine an organization's current position on the journey of becoming business process oriented and in developing a strategy for building the e-networks to compete in today's Internet economy. Understanding exactly where efforts should be focused and having a tool to measure progress should be valuable to companies involved in building business process orientation. In addition to the BPO measurement tool and maturity model, we have included an additional assessment tool in Appendix B. This tool, the Business Process Assessment Tool, provides an additional dimension to understanding the practices that are needed to be in place for successful BPO.

Building e-networks for today's Internet economy is a tremendous challenge. Building these networks at the process level is very dangerous without understanding the process orientation of each actor in the net-work. The problem of BPO incompatibility can destroy a network very quickly.

Chapter 6 will build on the use of the BPO tool and the BPO maturity framework by reviewing a case example of a major manufacturing com-pany's BPO journey.

NOTES

[1] Kotler, Philip, *Marketing Management*, Millennium Edition, Upper Saddle River, NJ, Prentice–Hall, 2000, p. 13.

6

INTRODUCING BPO IN MANUFACTURING

This chapter discusses the use of the BPO Assessment Tool and Maturity Model in examining the progress of a BPO implementation effort in a manufacturing setting. This effort was conducted within a major multinational pharmaceutical and chemical company that involved several divisions, several years, and several million dollars. The company involved also made a significant investment in enterprise technology, SAP, which was used as the catalyst and enabler for the organization to become more business process oriented.

TECHNOLOGY DRIVEN BPO IN MANUFACTURING

Many multinational manufacturing firms in the 1990s invested millions in the effort to improve performance though the use of technology-enabled change. The intent of many of these efforts was to improve internal operating efficiencies and reduce costs, by focusing more on the horizontal and becoming more business process oriented. Centralized data, common redesigned processes, and horizontal organization structures were the tactics used in many cases.

Many technology or software products were sold to these firms as vehicles for change. Consultants also sold this strategy (becoming more business process oriented though a technology-driven project) as the way to radically improve business performance. SAP, a German enterprise resource planning software firm, is the leader in this approach. The sales presentations for SAP and its consulting partners presented the objectives of BPO and the business process best practices incorporated in the SAP software.

From 1995 to 2000, many companies invested significant resources (some invested $100 million or more) to implement technology-driven BPO. The goals and strategies of these efforts were usually financially and systems project driven and measured. Project milestones, dollars spent, or number of modules installed were used as success measures. What it meant to be "horizontal" or business process oriented was usually undefined and thus difficult to measure.

With this information as background, the following sections discuss the efforts of a specific manufacturing company, Worldwide Laboratories, to become more business process oriented. The BPO measurement tool (Chapter 4) and maturity model (Chapter 5) were used toward the end of this multi-year effort to assess BPO progress.

Background of Worldwide Laboratories

Worldwide Laboratories is a very successful, well-known company in the pharmaceutical and chemical business. The overall holding company has several divisions, more than 130,000 employees, and more than $30 billion in annual global sales. The division undertaking this BPO effort was the U.S.-based group, with approximately 20,000 employees and $8 billion in sales per year.

The effort to implement SAP and become more business process oriented began in 1997 and included a series of projects organized by product line. The focus of the projects was to design and implement the new technology-based horizontal processes and, through that effort, to redesign jobs and structures that match these new horizontal processes. A significant effort was made in implementing a process-oriented measures and reporting system as well. Horizontal measures such as forecast accuracy, vender delivery performance, on-time shipments, and inventory were instituted, as were clear responsibilities for performance. Broad teams were organized across the business, specifically in supply chain and operations planning, to allocate resources, make decisions, and lead the activities needed to achieve the process performance goals.

The project was implemented by building execution teams representing most aspects of the business and empowering these teams to design, build, and implement the new system complete with new business processes, new jobs, and new measures and management systems. In 1999, these teams included over 300 Worldwide Laboratories employees involved in this effort on a full-time basis.

The results of this effort, as measured in early 1999, show significant return. Capacity was increased by 30% without any capital investment in machinery and equipment. Total inventory was reduced by 50% and the cash flow cycle time (the time from order receipt to the receipt of cash) was also reduced by 50%. All of these are remarkable business performance measures but how far had they come on their goal of becoming more business process oriented?

Charting BPO Progress

How far had Worldwide Laboratories progressed toward its goal of becoming more business process oriented? What areas are ahead and what areas are behind? What has been the impact on the organization?

The BPO measurement tool was used in order to answer these questions. The first step was to identify a list of "key informants" that could complete the survey questions. These Worldwide Laboratories employees selected had to represent the population of Worldwide Laboratories as a whole and could not be randomly selected. Therefore, the list of participants was developed by one of the key project leaders. As Figures 6.1 and 6.2 show, the respondents represented a broad cross section of functions and levels in the organization.

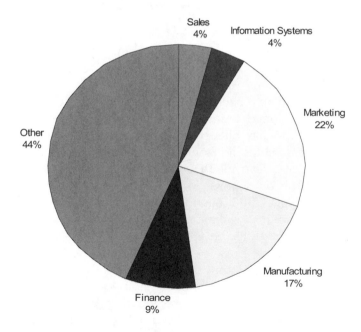

Figure 6.1 Functions Participating in the BPO Assessment

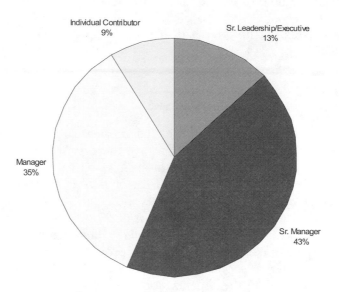

Figure 6.2 Job Positions Participating in the Assessment

There is an interesting observation with regard to the functions represented in the assessment. Almost 42% of the employees surveyed identified themselves as being from other than the traditional vertical functions (sales, manufacturing, marketing, etc.). Where questioned, most of these people had been assigned new broader process-oriented jobs during this redesign phase, and they no longer fit into the traditional functions or departments. In our research, this has often been a clear indication of moving to a structure that includes the horizontal or BPO aspect.

To determine the level of BPO maturity of Worldwide Laboratories, responses from the questionnaires were analyzed and the data plotted first on the high level BPO maturity model, then on the detailed BPO maturity model, and finally, compared to the specific answers in the BPO benchmarking database. Figure 6.3 shows the level of BPO maturity for Worldwide Laboratories.

Worldwide Laboratories appears only slightly above the Defined level. After several years of effort, tens of millions of dollars and over 300 people working on this project, it was expected that Worldwide Laboratories would be in the Linked level, rapidly moving into the Integrated level. This was a surprising finding and one that needed more examination.

The next step in the assessment involved the construction of the detailed BPO maturity model shown in Figure 6.4. This was constructed by calculating the totals for each category of questions and plotting the results on the model against the benchmarking data in the database.

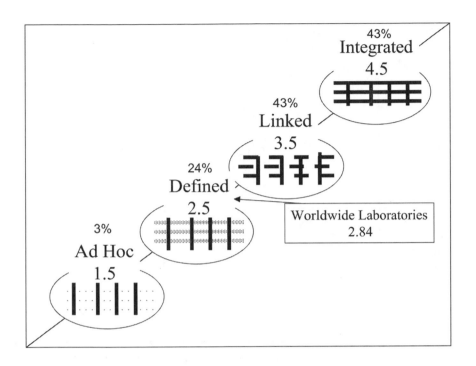

Figure 6.3 High Level BPO Maturity for Worldwide Laboratories

The detailed BPO maturity model shows clearly that Worldwide Laboratories is significantly into the Linked level in regard to PM. This would make sense since a strong focus of the project is to implement shared horizontal measures. But Worldwide Laboratories' rating in this area (13.73) is still below the database mean of 16.3. Why is this? One reason might be that the measures are not deep enough. They may be used only by the project teams and the leadership level of the organization for reporting not taking action or allocating resources. We will answer this question when we go more deeply into the data by comparing the answers for Worldwide Laboratories to the database benchmarks.

Another observation from the detailed maturity model is that the PV component is surprisingly low. Worldwide Laboratories' rating of 8.96 is below the mean and just barely out of the Ad Hoc maturity level. For a company that has spent several years and millions of dollars with a goal of documenting and redesigning common processes, this seems very low. One possible explanation that will be investigated in the detailed benchmarking is that the process documentation may have been constructed by the project team and remained within the team and not shared with the organization. This is often the case with technology-driven BPO efforts.

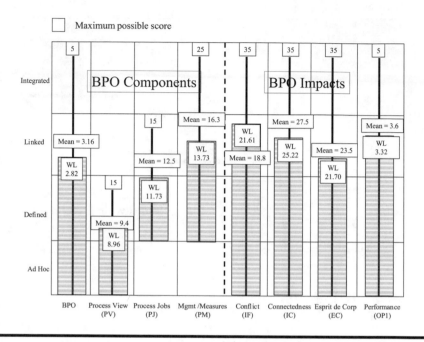

Figure 6.4 Detail BPO Maturity for Worldwide Laboratories

The teams view the process designs as engineering documents not as change tools to influence the orientation of the organization.

Another explanation for this low PV rating might be that the process view efforts were only focused on the processes related to the technology systems being implemented. If this were the case, this would only involve 20% or so of the total processes in a business and would clearly keep the PV rating below completely defined. Both of these avenues will be explored when we examine the detailed benchmarking data.

One last observation from the detailed maturity model is that the PJ component is below the mean (11.73 vs. 12.5). This again might be that the creation of the process-oriented jobs remains at the management level and was not really created deeply and broadly across the organization. This often happens in a BPO effort. The functions remain intact below the leadership level and BPO barely exists. BPO is then viewed as just another reorganization at the top rather than an organization-wide change in the way things are done.

All of these questions will be more deeply explored when the detailed answers are compared to the benchmarking database.

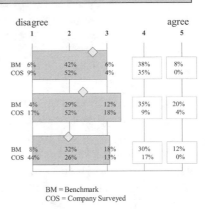

Figure 6.5 Answers to Detailed Process View Questions vs. Database Answers

Worldwide Laboratories Detailed Benchmarking Analysis

Figure 6.5 contains the detailed answers of Worldwide Laboratories compared to the benchmarking database. The diamond represents the mean of Worldwide Laboratories answers for that question and the horizontal bar shows the mean answer in the database. The percentages listed for each numerical answer, 1 through 5, on the line marked BM represent the percentage of survey responses in the database that answered the questions with that number. Worldwide Laboratories' percentages are listed right below the BM line. For example, for question PV1, which indicates that the average employee views the business as a series of linked processes, 52% of the people surveyed in Worldwide Laboratories answered this question with a 2 ("mostly disagree"). Thirty-five percent of the people answered this question with a 4 ("mostly agree"). This matches, almost exactly, what we see in the database (2 to 42% and 4 to 38%).

Prescription for Process Improvement

This shows a large and clear disagreement and would support the earlier diagnoses that the project team and the leadership involved in the project may have a process view but it is not broadly accepted within the organization. This has been a common problem with technology-led BPO efforts. A further sort of the data by job position supported this theory. The senior manager and leadership category generally answered "mostly agree" to this question while the individual contributor category answered "mostly disagree."

The action needed to correct this imbalance was to undertake "validation and involvement" activities broadly and deeply across the organization. This would involve cross-functional sessions of individual contributors and senior managers where the project team offers the "view" of the business processes and the group would discuss, possibly modify, and accept.

This activity is time consuming and expensive, which is why many organizations avoid it. But in our research, acceptance by all members of the organization is critical to building a business process orientation and can only be done individually.

Worldwide Laboratories' answers for question PV3, "The business processes are sufficiently defined so that most people in the organization know how they work," seems to support this diagnosis. Forty-four percent completely disagree with this statement and feel that the processes are not sufficiently defined or broadly understood. In the benchmarking database only 8% answer this question "completely disagree." This indicates support for the earlier recommendation and suggests that the current documentation may not be sufficient. It may not include the non-technology-related process activities or it may lack sufficient detail in the non-technology-associated processes.

In fact, upon further examination of the documentation, both of these were the case. In order to keep to the project schedule and minimize costs, the project team had included mostly the technology-related process activities (estimated to be less that 30% of the business processes). They had barely touched on the non-technology-related activities even within the business processes dramatically impacted by the technology project.

Question PV2, "Process terms such as input, output, process and process owners are used in conversation in the organization," is meant to measure the level of process language adoption. From our research, the acceptance and use of a process language in an organization is a key factor in implementing BPO. The leading firms in our database show up as having answered "mostly agree" and "completely agree" at 20 and 30% levels.

Worldwide Laboratories clearly has a problem in the area that supports the earlier diagnosis and recommendation: 69% of the respondents answered "completely disagree" or "mostly disagree." This shows almost a total lack of process language in use. Building this can only be an individual effort and is a major barrier to many companies trying to build BPO. This again is expensive and time consuming but a critical foundation for moving out of the Ad Hoc level of maturity.

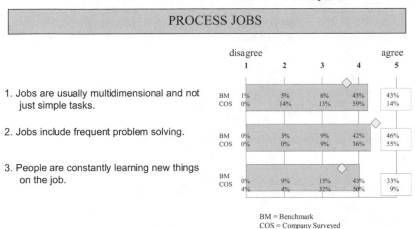

Figure 6.6 Answers to Detailed Process Jobs Questions vs. Database Answers

Figure 6.6 shows the Worldwide Laboratories answers regarding process jobs (PJ). It is clear that the company has achieved results in this area. The multi-dimensionality, problem-solving, and learning aspects are all showing up with the majority of answers at the "mostly agree" and "completely agree" levels.

Sustaining Process Change

The sustaining process change has been an area of concentrated effort in the U.S. for several years and is showing results. This is probably a big factor in the performance results achieved by Worldwide Laboratories. Unfortunately, unless the process view issues are addressed, significant organization stress will result. Battles between cross-functional, process-oriented jobs, and formalized functional power centers will result. Since the process boundaries are not defined and accepted, decisions will be made and resources will be allocated based upon functional authority, since it is the only one defined in the budgeting and management areas of the business. A PJ without solid authority will soon crumble. This area may now be a strength for Worldwide Laboratories, but without the support process view there is a strong risk that this area will be a powder keg that will degrade back to functional jobs. An analogy for this would be a highly trained driver who is given a high-powered car without a good chassis. Disaster will eventually result.

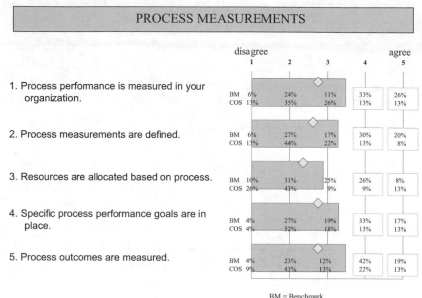

Business Process Orientation Survey

PROCESS MEASUREMENTS

	disagree 1	2	3	4	agree 5
1. Process performance is measured in your organization.	BM 6% COS 13%	24% 35%	11% 26%	33% 13%	26% 13%
2. Process measurements are defined.	BM 6% COS 13%	27% 44%	17% 22%	30% 13%	20% 8%
3. Resources are allocated based on process.	BM 10% COS 26%	31% 43%	25% 9%	26% 9%	8% 13%
4. Specific process performance goals are in place.	BM 4% COS 4%	27% 52%	19% 18%	33% 13%	17% 13%
5. Process outcomes are measured.	BM 4% COS 9%	23% 43%	12% 13%	42% 22%	19% 13%

BM = Benchmark
COS = Company Surveyed

Figure 6.7 Answers to Detailed Process Management and Measures Questions vs. Database Answers

Figure 6.7, the final BPO component to be examined, represents the area of PM and has been a major focus for Worldwide Laboratories' project. This being the case, the answers are a bit disappointing. The answers to PM1, "Process performance is measured in your organization," clearly indicate that the vast majority of people surveyed believe it is not measured (over 70% answer 3 or lower). This suggests that the measure implemented is not broadly communicated or used and probably utilized only as management reporting tools. When we sorted this question by position in the organization, it became clear that senior leadership indicated that process performance was measured and the remainder of the organization responded that it was not.

The answers to PM2, "Process measurements are defined," clearly indicates that the process measures were not included in the project effort, or they were not communicated (44% "somewhat disagree"). This again is a common problem for technology-driven BPO efforts. Deep process measures are not needed to implement an information system so the team does not focus on them.

A clear solution to this is to build the cascading PM system required to focus people at all levels on the horizontal, process performance. There is a clear separation in our database between the people that answer 4 and 5 to this question ("mostly agree" and "completely agree") and the people that answer 3 or below. PM is the BPO component that is the most strongly linked to EC, conflict, connectedness, and overall performance. The organizations that show a score of 4 or 5 on all three PM questions clearly outperform organizations that do not. Management reporting systems do not orient an organization toward process; they orient an organization toward the hierarchy and reinforce the functional structure.

A key question describing this is question PM3, "Resources are allocated based upon process." This is the fuel that makes an organization run. A PJ without resources aligned and available to focus on the process will surely fail. This question was answered as "mostly disagree" or "completely disagree" by 68% of the Worldwide Laboratories respondents. This means that the functions are still predominant when it comes to resource allocation. Budgeting, staffing, and attention of management, a scarce resource not often discussed, are still functionally oriented.

Outcomes of BPO on Worldwide Laboratories

As we examine the BPO impacts section of Worldwide Laboratories' assessment we can see validation of the earlier diagnosis of compartmentalization of the BPO effort at the top and in the project team. Some IF (interfunctional conflict) questions show the results of measurement misalignment. IF4 for example, "Employees from different departments feel that the goals of their respective departments are in harmony with each other," was answered "mostly disagree" by 55% of the respondents and "mostly agree" by 27%. When examined further, the break between senior leadership and the project team with the remaining parts of the organization was clear. This supports the importance of addressing the process measures and management issue.

The IC (interdepartmental connectedness) results indicate a strong horizontal connection among people. This supports the strong results shown in the process jobs category and the fact that 42% of the respondents had jobs outside the normal functional classification. Individuals, for short periods, can often overcome the weaknesses of structure. This "hero" activity has solved a lot of problems in the short term but has not been sustainable. Without organization alignment and authority, the functional conflicts soon bring the heroes down since they directly challenge the status quo. Charismatic leaders have relied on this approach for many years but when the leader goes, the heroes are shot and the status quo takes over. This "rubber band" effect is well known and has been the subject of many business books.

Esprit de corps (EC) for Worldwide Laboratories is lower than the database mean. As we look at the details it becomes clear that many people do not display "enthusiasm," a necessary ingredient of EC. The majority of respondents answered 3 (neither agree nor disagree) or below to five out of seven of the EC questions. This can be a result of many factors but is clearly not the engine needed in today's economy.

The process jobs without resources, the compartmentalized measures, and the lack of a broad process view can all impact this. The inherent conflict between functions and process has not been put in balance across the organization. It has begun but many people are on the fence with a "wait and see" attitude. In order to get to the Linked level or beyond, this impact area must be much higher.

CONCLUSION

Worldwide Laboratories has spent a tremendous amount of time, money, and organizational energy on its effort to become more business process oriented but this is not reflected in the results. Concentrating on completing a technology project at the expense of actually changing the orientation of an organization broadly and deeply has given the company the illusion of success but not a sustainable change.

The heroes created by this project have delivered short-term financial results but have not really delivered BPO. Cross-functional jobs without the authority or resources to sustain the effort past individuals are a "work around" and will crumble. This is shown by the weakness in esprit de corps, the engine of structural enthusiasm.

Before Worldwide Laboratories attempts to connect outward to an e-business network it must go back and build a broad and deep process view through validation and involvement within the organization. Cascading process measures must also be put in place to horizontally align the entire organization. The management reporting system implemented by the project and currently in place at Worldwide Laboratories only reinforces the functional orientation.

Worldwide Laboratories may have successfully implemented a systems project that produced some short-term financial benefits but it has not become business process oriented and has not built something sustainable that will function much beyond the end of the project. A foundation is critical to BPO just as it is to any effort. Worldwide Laboratories has a very weak foundation and is in danger of collapse. For more information on building this foundation in a manufacturing company, refer to the case study on New South, Inc. in Appendix A. Chapter 7 will discuss applying BPO to a service organization.

7

APPLYING BPO TO SERVICE OPERATIONS

This chapter covers the use of the BPO Assessment Tool and Maturity Model in examining the level of BPO in an information technology service organization that had experienced significant merger activity and technology change. We will discuss the unique aspects of BPO within a service organization and address the areas of greatest opportunity for improved performance.

IMPORTANCE OF PROCESS ORIENTATION IN SERVICE BUSINESSES

Low levels of productivity in the service economy have been a cause for concern among economic and business commentators for at least a decade. By 1995, private and public service industries accounted for 75% of Gross Domestic Product (GDP) and 78% of employment. With the service sector accounting for such a substantial part of all economic activity and employment, it has been noted with increasing alarm that service industries have not followed manufacturers in realizing steady gains in labor productivity.[1] The dynamics of a service company appear to be very different from those of a manufacturing company. The "product" in a service company is constructed and delivered right in front of the customer. This construction and delivery is almost totally dependent upon the company employee's attitude, skill, and performance. For example, a washer/dryer repair service company's product is the repairman at your house working on your washer or dryer. He designs and develops the "product" based upon his diagnoses of your problems and delivers it right there on the spot. If the repairman has poor diagnostic skills, your "product" may not work and the "consumption" of your product may be a very unpleasant experience when you try to use the machine. A repairman's rudeness only adds to the poor quality of the experience.

A manufacturing company's product, on the other hand, is designed and constructed based upon the assumption of the customers' needs and when they need it. Due to high levels of investments in automation, the construction of the product is often highly machine dependent and not as dependent on individual performance, which is the case in services. Often, the delivery and consumption of a manufactured product is decoupled, both in time and place, from the design and construction. From our research, we believe that, due to these differences, BPO has a different impact on a service organization's performance.

Given this context, this chapter investigates how BPO affects organizational culture and performance in a service business, by reporting research we recently conducted on the rapidly growing informational technology (IT) sector. During the 1990s, the information technology services market, growing at a consistent 20 to 30% annual growth rate, underwent massive changes. Not only did several major technology shifts occur (i.e., mainframe to client server to Internet computing paradigms), the industry underwent massive start-ups, mergers, and consolidations. Toward the end of the decade, capital was readily available from venture capitalists and through initial public offerings (IPOs). Acquisitions funded totally through stock swaps were cashless transactions that required very little funding. All of this led to small companies acquiring larger companies and regional companies becoming national companies overnight.

Many smaller firms had expended considerable effort building the culture and infrastructure needed to be a successful service company. Organizations were very lean, flat, and flexible. Simple, logical processes were quickly designed and implemented with minimal investment. Since these companies were focused upon a few local relationships with customers in certain geographic areas, they were also very close to the customer. Many of these firms were also employee owned, which resulted in a strong, natural sense of esprit de corps.

This merger and consolidation hyperactivity had a major impact on the people, structure, and processes of the affected companies. Employee turnover after an acquisition became a major issue, as well as merging support processes and technologies.

With this as background, the remainder of this chapter discusses a specific IT service company case (referred to as CL Technologies) and the impact of turbulent change on its processes and people as measured using the BPO questionnaire (see Appendix C). We will also explore the different BPO relationships in a service company as opposed to a manufacturing company.

Background of CL Technologies

CL Technologies provided IT consulting, design, and implementation services to a broad cross section of industries (banking, manufacturing, insurance, health care) and was the result of a single firm acquiring three additional firms within 2 years. Each of the acquired firms had approximately $100 million in annual revenues with 700 to 800 employees. As with many service companies, most of the billable work done by the acquired firms was performed on a customer's site. Small offices within each acquired company had been formed around successful client engagements. Many of the acquired firms' offices, which numbered 30 or more, were small groups of less than 75 people. Any corporate management consisted of a very small team of executives rather than any corporate office. This is very consistent with a service company's need to maintain its "inventory" (the people delivering the service) close to the customer. The local offices were more like families than work places. The acquiring firm was quite different with a hierarchical approach and one large office in a major U.S. city and one office outside the United States.

The integration approach used to bring all the companies together consisted of teams from the acquired companies being formed with a leader from the acquiring firm. The teams focused on building the common process to be implemented for the new company. This integration strategy consisted of the teams examining the processes and practices for each company and selecting the best ones. If none of the processes fit, the team would be asked to design and build one for the new company. This integration process was expected to take only a few months.

A clash of cultures between the acquiring firm and the acquired firms presented significant obstacles. The acquiring firm believed in centralized, functionally oriented command-and-control processes and procedures while the acquired firms believed in exactly the opposite: decentralized horizontal processes with distributed authority. This argument dominated the integration discussions.

BPO in a Service Company and the Impact of Process Change

What are the dynamics of BPO in a service company? What is the impact of all this change on BPO and the organizational outcomes?

In an attempt to answer these questions, the BPO measurement tool was used. The first step, as with the manufacturing company, Worldwide Laboratories, discussed in Chapter 6, was to identify a list of "key informants" who could complete the survey questions. CL Technologies' key informant employees were selected based upon title and location in an attempt to represent the population of CL Technologies as a whole. Figures 7.1 and 7.2 show the distribution of the respondents. A broad cross section

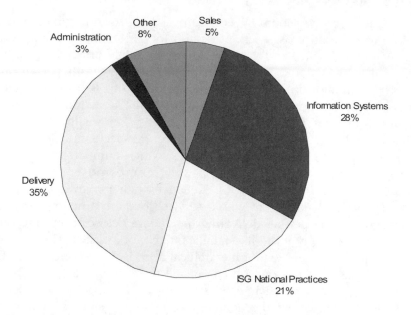

Figure 7.1 Functions Participating in the BPO Assessment

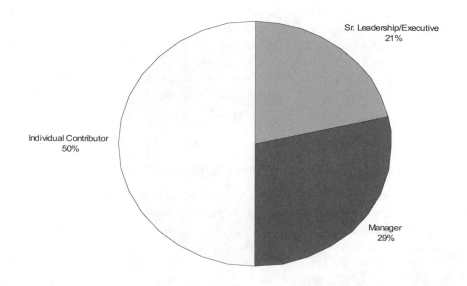

Figure 7.2 Job Positions Participating in the Assessment

of functions and levels in the organization was represented in the study and should be reflective of the entire organization.

CL Technologies had a simple and naturally process-oriented approach to organizing. As is the case with most service companies, delivery and support represented the core processes. A person was either in delivery, support, or a special "expert" group of highly paid, hard-to-find people who traveled to the client's site. IT services, by nature, are performed on the client's site; therefore, delivery was local while the several competency-focused teams, such as strategic consulting and certain technology experts, were organized nationally. Each office also had small local support staffs and processes that fed into the overall corporate support organizations and processes.

The delivery of the IT service was performed on a project-by-project basis. A client's needs would be identified, and a project team would be formed around developing, designing, and constructing a solution to fit that need. As this delivery team needed resources, it would request them from the local office or the national team of experts. When the client was satisfied, the team would disband, go back into the resource pool, and be reassigned to another project.

In order to examine the level of BPO maturity of CL Technologies, responses from the BPO questionnaires were analyzed and the data plotted. First, the high level BPO maturity was calculated, then examined on the detailed BPO maturity model, and finally compared to the specific answers in the BPO benchmarking database. Figure 7.3 shows the level of BPO maturity for CL Technologies.

CL Technologies appears slightly into the Linked level at a BPO rating of 2.92 out of 5. The BPO assessment tool was not used before the mergers. Based on several interviews, we concluded that all firms were significantly into the Linked level prior to the mergers. When interviewed, people from each company described efficient, smoothly operating pre-merger processes that were well documented and understood by the organization. Teams were organized by process and process owners. Measures were in place in many cases. In fact, one firm was ISO certified, which means process documentation is substantially in place and maintained.

With the massive disruption from mergers, technology change, and process integration efforts challenged by significant conflicts of operating philosophy, this BPO result of 2.92 is very surprising. Based upon past experience with manufacturing companies, we expected a more negative impact on BPO and a score more toward the Ad Hoc level of 1.5.

To fully examine this surprising result, we plotted CL Technologies' results on the detailed BPO maturity model shown in Figure 7.4. This was constructed by calculating the totals for each category of questions and

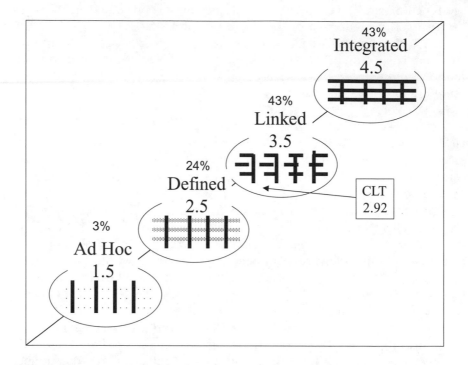

Figure 7.3 High Level BPO Maturity for CL Technologies

plotting the results on the model against the benchmarking data in the database.

The detailed BPO maturity model shows clearly that CL Technologies is into the Linked level in PM and just barely into the Defined stage with PV. The PV area has significantly degraded due to the process integration efforts, disruptions, and mergers, and is about where we had expected, close to the Ad Hoc level but still into Defined. The PJ and measures category seems to indicate a strong foundation for moving into the Linked level if the PV issues can be addressed.

This is an interesting disconnect. PJ and PV generally track together. Companies whose processes are not properly designed or documented are, by definition, Ad Hoc. Jobs and measures are difficult to build around undefined processes. Our preliminary conclusion for this particular service enterprise was that, short of changing your customers and their requirements, core processes stay the same. Therefore, what happened to the process documentation and people's understanding of the processes? In a service company, they are imbedded in the customer's situation and the service person's skill and experience. Upon questioning of the participants, the PV results seem to only reflect the status of the integration teams'

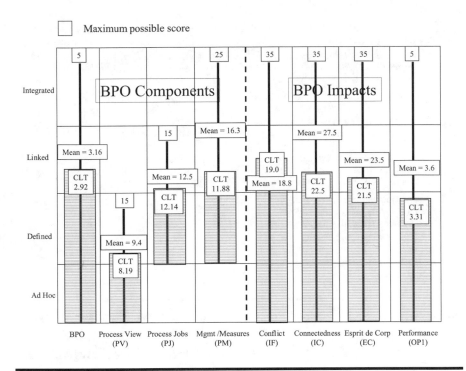

Figure 7.4 Detailed BPO Maturity for CL Technologies

progress on discussion, agreement, and documentation around common support processes, not the core service delivery process.

Conflict, which is a reversed scale on the detailed maturity model, shows that CL Technologies is above the mean (19.0) and has slightly lower conflict than the mean in the database (18.8). This is surprising considering the operating philosophy conflicts brought out by the interviews. Connectedness (IC) and EC are both below the mean, as was expected during times of disruption, but are still surprisingly at the Linked level. These results are close, and in some areas, higher than Worldwide Industries, the manufacturing company examined in Chapter 6. Worldwide Industries spent tens of millions of dollars and several years on trying to improve its BPO. CL Technologies, on the other hand, spent almost nothing and had, in fact, experienced more disruption in 1 month than most firms experience in several years. This seems to indicate that a service business, or at least an IT service, is by nature, business process oriented.

In the next section, we will dig more deeply into this aspect by reviewing the detailed results from CL Technologies compared with the database benchmarks.

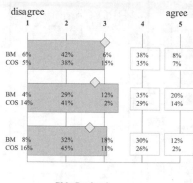

Business Process Orientation Survey

PROCESS VIEW (PV)

Figure 7.5 Answers to Detailed Process View Questions vs. Database Answers

CL TECHNOLOGIES DETAILED BENCHMARKING ANALYSIS

Figure 7.5 contains the detailed answers of CL Technologies compared to the benchmarking database. As in the Chapter 6 analysis, the diamond represents the mean of CL Technologies answers for that question and the horizontal bar shows the mean answer in the database. Because means or averages can sometimes hide the details, we have included the distribution of the respondents' answers. The percentages listed for each numerical answer, 1 through 5, on the line marked BM represents the percentage of survey responses in the database that answered the questions with that number. CL Technologies percentages are listed right below the BM line on the line marked COS. For example, for question PV1 indicating that the average employee views the business as a series of linked processes, 38% of the people surveyed in CL Technologies answered this question with a 2, "mostly disagree." Thirty-five percent of the people answered this question with a 4, "mostly agree." This matches almost exactly with what we see in the database (2 to 42% and 4 to 38%).

DIAGNOSING THE ISSUES

Looking past the mean at the distribution of the answers, we see that in each question shown in Figure 7.5 there are two groups of responses concerning the BPO component, process view (PV). One group answered a 2, disagreeing with the statements and one group answered 4 or 5, agreeing with the statement. This is obviously an area of disagreement within the organization. How is this possible?

To answer this question we examined the data by position in the organization. For PV1, "The average employee views the business as a series of linked processes," senior leadership and some managers accounted for most of the 4 answers, "mostly agree," while individual contributors accounted for most of the 2 answers, "mostly disagree." This can be explained by the fact that the individual contributors were decentralized because of their delivery role. They generally worked on client sites and smaller offices while the senior leadership and managers work in the headquarters offices. Since the acquiring company's philosophy for centralized processes was dominant, they were building these centralized processes. The decentralized processes known to the individual contributors were being eliminated and consolidated into the headquarters' centralized, functionally oriented processes, and the decentralized people, the individual contributors, were not involved in the new process. Therefore, the individual contributors in the delivery area believed that the organization did not have a very mature process view while the people in headquarters building the new centralized processes believed that they did have a mature process view.

Also, since the delivery processes of an IT service company are local because of the nature of the business, they are difficult, if not impossible to centralize. Therefore, the processes being referenced by the respondents to this question were only the support processes. The individual contributors seem to view their personal delivery processes as their own, not the processes of the company, and therefore excluded them for consideration in the assessment of process view.

PV3, "business processes are sufficiently defined" and "most people in the organization know how they work," confirms this conclusion. There is a clear break between the 61% (people who answered 1 or 2) who said "no" to that question and the 28% (people who answered 4 or 5) who said "yes." The headquarters groups building the centralized processes believe that the processes are defined and understood by the organization because the people building and running the centralized processes do understand them. Unfortunately, the people supported by and interacting with these processes say that this is not true. PV2 shows the same pattern.

The detailed examination of these results and follow-up interviews provided some interesting perspectives. We found that the only processes being considered during the BPO assessment by the people working in the delivery processes were the company support processes. They considered the delivery process to be their own and affirmed that this is a process they definitely understood. As in the earlier example with World-wide Industries, building and running centralized processes without deep involvement of all or most employees does not make an organization process oriented. To have a process view, the organization and the employees must be involved and understand the process and use a common process language. (Refer to the ABIG Case in Appendix A for another perspective on process change in a service business.) In the case of CL Technologies, the new centralized support processes were created and operated in such a way as to degrade the process maturity, at least in the PV category.

Since BPO, and particularly the PV component, is related to overall business performance and EC, will perceived ownership of these support processes by employees affect business performance and EC? From our research, it appears that in an IT service company the impact from PV on EC and overall performance is not that strong. CL Technologies' EC level is close to the mean and very high for a situation with this much turbulence and change. Figure 7.6 shows the detailed results of the survey.

When we drilled down into the data, it became clear that the people working within the delivery processes had a naturally high level of EC because of the nature of the IT service being performed. Those individuals from the corporate office, however, who were trying to centralize control, answered very negatively. For example, question 2 asks if team spirit pervades all ranks of this business unit. Fifty-five percent of the employees, mostly the decentralized delivery people working at client, answered "yes" while 26%, mostly the corporate staff, answered "no." Is there a logical explanation for this apparent discrepancy? Our opinion is that the IT business, by its very nature, should have decentralized processes in order to take advantage of the natural level of EC.

Alignment of process strategies with the dynamics of a market is important because organizational energy and success comes from it. A market such as the IT services, which by nature requires decentralized processes, will not react well to a centralized process mentality. We believe that business performance, as well as esprit de corps will suffer. If the processes are not aligned with the natural aspects of a market, then inevitable conflicts will subtract from the natural feelings of esprit de corps in a decentralized IT service business. Where there is alignment with market forces, then conflict and connectedness to the market will positively impact process performance and esprit de corps. This "process synergy" can be a powerful force that can energize an organization.

In order to explore the nature of the service business further, the PJ component should be discussed. It is already clear from the earlier discussions that, in an IT service business, people view processes differently than in a manufacturing company. Since PJ exist within a process context, this decentralized view must also exist.

Business Process Orientation Survey

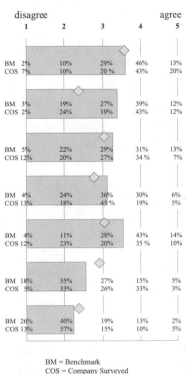

ESPRIT DE CORPS (EC)

	disagree			agree
	1	2	3	4	5

1. People in this business unit are genuinely concerned about the needs and problems of each other.

	1	2	3	4	5
BM	2%	10%	29%	46%	13%
COS	7%	10%	20%	43%	20%

2. A team spirit pervades all ranks in this business unit.

BM	3%	19%	27%	39%	12%
COS	2%	24%	19%	43%	12%

3. Working for this business unit is like being part of a family.

BM	5%	22%	29%	31%	13%
COS	12%	20%	27%	34%	7%

4. People in this business unit feel emotionally attached to each other.

BM	4%	24%	36%	30%	6%
COS	13%	18%	45%	19%	5%

5. People in this business unit feel like they are "in it together."

BM	4%	11%	28%	43%	14%
COS	12%	23%	20%	35%	10%

6. This business unit lacks an "esprit de corps."

BM	18%	35%	27%	15%	5%
COS	5%	33%	26%	33%	3%

7. People in this business unit view themselves as independent individuals who have to tolerate others around them.

BM	26%	40%	19%	13%	2%
COS	13%	57%	15%	10%	5%

BM = Benchmark
COS = Company Surveyed

Figure 7.6 Answers to Detailed Esprit De Corps Questions vs. Database Answers

Figure 7.7 shows CL Technologies' answers regarding PJ. For questions PJ1 through PJ3, 70 to 90% of the people mostly or completely agree. Each of these questions reflects the multidimensional, problem solving, and learning aspect of the jobs. When we examined this further, it made sense and agreed with our earlier assessment. The nature of the IT service industry and many other service industries is that the "product" is in reality the individual performing the service with the client. By definition, this places the final responsibility for process performance, learning, and continuous improvement on the individual performing the service. This confirms our earlier conclusion that process decentralization naturally forms in IT service businesses which are driven by constant change and are almost impossible to centralize. It is no surprise, then, that PJ1 and 2 are so high.

Business Process Orientation Survey

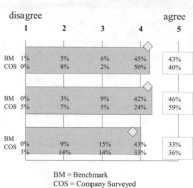

| | disagree | | | | agree |

PROCESS JOBS

1. Jobs are usually multidimensional and not just simple tasks.

2. Jobs include frequent problem solving.

3. People are constantly learning new things on the job.

BM = Benchmark
COS = Company Surveyed

Figure 7.7 Answers to Detailed Process Jobs Questions vs. Database Answers

PJ3, "learning new things on the job," has a different dynamic. In the IT service industry, which has been rocked by technology shifts, learning new things on the job has become essential to survival. This high rating is no surprise. In fact, most industries in our database have a high rating in this area. We believe it is due to the major changes currently taking place. New technologies, reorganization, downsizing, and 20 years of Total quality management (TQM) have all impacted this BPO component.

We believe that this is a natural dynamic of the services business in general. It follows that if new products are developed by "integrating the voice of the customer," then service development and improvement must be highly spontaneous and adaptive, sometimes even while the service is being delivered. Thus, service businesses and their employees must be constantly learning and adapting to new service encounters.

Aligning and Focusing the Organization

Process measures focus and align the people in an organization toward the appropriate outcomes, assuming, of course, the measures are correct. When the process measures collapse, significant performance decrease soon follows. Confusion and frustration also follow, thus increasing conflict.

Figure 7.8 is the final CL Technologies BPO component to be examined. There are major issues in this area. For each question, 50 to 70% of the people answered mostly or completely disagree. Our conclusion is that

Business Process Orientation Survey

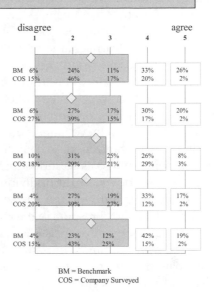

PROCESS MEASUREMENTS

	disagree			agree	
	1	2	3	4	5
1. Process performance is measured in your organization.	BM 6%	24%	11%	33%	26%
	COS 15%	46%	17%	20%	2%
2. Process measurements are defined.	BM 6%	27%	17%	30%	20%
	COS 27%	39%	15%	17%	2%
3. Resources are allocated based on process.	BM 10%	31%	25%	26%	8%
	COS 18%	29%	21%	29%	3%
4. Specific process performance goals are in place.	BM 4%	27%	19%	33%	17%
	COS 20%	39%	27%	12%	2%
5. Process outcomes are measured.	BM 4%	23%	12%	42%	19%
	COS 15%	43%	25%	15%	2%

BM = Benchmark
COS = Company Surveyed

Figure 7.8 Answers to Detailed Process Management and Measures Questions vs. Database Answers

the leaders of CL Technologies, who have a naturally decentralized delivery model by the nature of the business, have no measures, and therefore, no effective way to see, understand, or influence what is going on within the organization. According to the data, process performance measures are not in place or defined, and goals are not in place. Process outcome measures are also not in place. How is process success measured and how do you know if you are going in the right direction?

We discovered through interviews that customer delivery measures are clear, in place, and used by the delivery teams working at client sites to manage the delivery processes. Since the services are being performed in direct contact with the customer, measurement and feedback of key aspects are easy and immediate. A problem occurs when the centralized managers want to see and control this process. Since information regarding the performance against these measures is staying with the delivery teams and not being shared, the centralized managers are "blind" with regard to process performance.

By examining the data and interviewing participants, we found that in the service business, most employees view measures as a management tool to check up on people. The delivery teams do not view the informal way they measure their own delivery as falling under the measures category since they do not share the results with anyone but themselves. These individualized, somewhat informal measures do not lead to uniform service across different business units or geographies. For example, clients may not be receiving the same thing from different delivery. Moreover, comparisons of learning among delivery teams is difficult unless consistent measures are used. Good process performance is subject to the interpretation of each person when individual measures are used. An organization working together to achieve market goals must have alignment of process practices, even in a decentralized service business.

In order to achieve alignment, CL Technologies needs to quickly involve the organization in building cascading measures that connect overall company goals to individual actions. This must be conducted for each process down to the individual delivery teams. With the naturally decentralized nature of the IT service business, only process control performed by those responsible for the process can work. Yet, conversely, uniform quality and consistency across many locations can only be achieved with common measures and aligned goals. Without measures and goals that support the direction of the company, each individual is an island and leadership is unable to understand what is going on, how to influence events, and where to invest resources.

The lack of measures might fuel the drive to centralize so that visibility and control can be achieved. Yet, this contradicts the natural decentralization of the IT services business and will impede flexibility and responsiveness. Considering the rapid changes of the technology business and the need for individual employee responsiveness, centralization will present a major barrier to success.

Outcomes of BPO on CL Technologies

With an organization going through such changes and potential conflicts, the BPO outcomes are surprisingly high. Forty to seventy percent of employees answered the esprit de corps questions positively (4 or 5). Conflict was surprisingly low and connectedness much higher than expected.

Why this unusually strong relationship between BPO and CL Technologies' organizational culture (e.g., EC, internal connectedness)? Perhaps this result is explained by the natural team oriented way IT services are organized and performed, which by definition forces decentralization. As is the case with most service businesses, delivery of the service was performed in direct contact with the customer at the customer's site and thus demanded decentralization. Project teams had to be formed and oriented around the customer and each was treated as its own "business unit." This resulted in each team being a self-supporting business unit focused on a customer. These small project teams were, by nature, very entrepreneurial and integrated. A feeling of esprit de corps and reduced conflict was a natural outgrowth and essential to survival.

CONCLUSION

It is clear from this research that a service business has very different BPO dynamics than a manufacturing company. The creation and delivery of services directly to a client is very different, demanding decentralized processes and jobs. The very survival of a service company depends on how well individual employees execute the service offering. In the IT service business, as well as many project-oriented businesses, this leads to small, tight groups that operate as independent business units. This service dynamic results in a natural process orientation, yielding greater esprit de corps, lower conflict, and greater connectedness.

CL Technologies, like most of the IT service business, has experienced significant disruption. In spite of this, CL Technologies seems to have maintained a natural connectedness, esprit de corps, and low conflict due to the decentralized, project team-oriented IT service business. Process jobs seem to be the natural and almost unavoidable way to operate. This indicates the strength of natural BPO in a service company.

To take advantage of this powerful BPO force that exists in an IT services company, the organization must be aligned with the natural patterns of the business (decentralization) rather that using tremendous energy (centralization) to fight against them. CL Technologies can dramatically improve its BPO maturity and performance and allow them to direct resources to the right activities by developing aligned, cascading process goals and measures. With this alignment will come higher levels of business performance and esprit de corps achieved with a much lower investment than Worldwide Industries. If CL Technology can achieve this level of BPO with almost no investment and in the middle of these negative forces, just imagine what can be achieved with alignment of the efforts of the organization.

In conclusion, the IT service business, and we feel most other service businesses, have a natural tendency toward BPO. By aligning a company's efforts with these natural market forces a service organization can more quickly reach higher levels of BPO and the resulting outcomes. With a manufacturing firm that lacks this natural BPO tendency (such as Worldwide Industries) significant investments are required to reach even the *Linked* level of BPO. CL Technologies has achieved this even with investments that contradict BPO and in the middle of major disruption caused by mergers and acquisitions. BPO appears to be a natural source of competitive advantage for service businesses to access, if they can only remove the barriers from conflicting management strategies.

How do these findings in the IT service business relate to other service markets? How do they relate to the new Internet economy? The natural market forces in a service business seem to demand process decentralization and process orientation. Healthcare, food service, and other areas clearly need process performance uniformity during each customer experience in order to be effective. Establishing a brand identity through consistent, high quality interaction with a customer is a hallmark of McDonalds. Part of its success is due to a high level of process orientation in their methods, jobs, and structures. We believe strong BPO is the key to any service business. The Boston Market case (see Appendix A) further examines process performance, uniformity, and an individual's role in achieving BPO.

What about the new Internet economy? We believe BPO is a key factor. The Internet allows personalized delivery of services in a global way. The concept of "place" on the Internet is no longer where the company is located but where the customer is interacting with a process on the Internet. Although the interactions might be occurring on the customer's site, the people doing the interacting can be anywhere.

At the same time, many service processes that once could only be delivered in person can now be digitized and available on the Web. One example is a travel agent. When considering a vacation, you can now quickly search for travel ideas and even "visit" the location through video tours of your hotel room and the facilities. Once you have decided on what you want to do you can post your request online and travel suppliers will bid on your business. Then you can make the purchase if the price is right. All this can be completed without leaving your home. Your travel agency consists of your Web search engine and your home computer and the collection of companies that you interacted with during your investigation and purchase.

What does this mean for process orientation? In order for all of these processes to be called upon at your request they must be combined at the process level. Common search terms or key words are a start. A common set of words to describe the offerings and a common price structure for comparisons are essential. These all come from developing a process view, one of the key BPO components.

What does this mean for the service organization? It means that, in the case of a travel-related business, process centralization must occur but employees can be decentralized. As long as the unified connection point is at the customer's computer (and phone) and not the employee's location, then the Internet allows total decentralization of a company's employees.

Just as in the IT services business, this puts the focus and responsibility on the individual for process performance, while recognizing a common, centralized delivery process. In a totally decentralized organization, connectedness and esprit de corps will be great challenges. BPO can possibly help these new e-services organizations grasp these complexities. Chapter 8 will discuss research investigating the application of BPO to the supply chains of organizations.

NOTE

[1] *Bureau of Economic Analysis*, U.S. Department of Labor, Survey of Current Business, 78, September 1998.

8

BPO AND SUPPLY CHAIN MANAGEMENT

This chapter reports on a research project that adapted the BPO measurement instrument to the Supply Chain Council's (a global organization formed to define and benchmark supply chains) model for describing and measuring supply chains. BPO-related questions were developed, pretested, and used to evaluate the impact of BPO on supply chain decision process performance, or supply chain management, in a cross-industry study. Conclusions from our study were quite clear: BPO is a major contributor to improving supply chain management performance.

WHAT IS THE SUPPLY CHAIN MANAGEMENT?

The major core business processes today, such as demand creation, demand conversion, product development (PDM), customer relationship management (CRM), and supply chain management (SCM), are being challenged to meet rising demands and expectations. However, the most critical business processes being challenged are those which contribute to the core value received by the customer. These processes take inputs from suppliers, add the strategic value, and deliver the final value package to the customer. Effective management of these processes is critical to any company.

Clearly, the era when the focus was on managing the supply chain of a single company is over. Today, these processes can and often do transcend company boundaries and involve cross-company planning and implementation. As Figure 8.1 shows, a company's application of supply chain management has evolved from designing and managing the supply chain to obtaining the functionally best supplies to design, manage, and

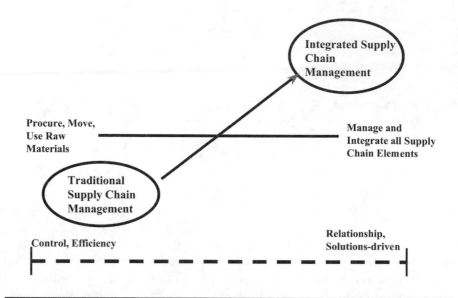

Figure 8.1 Traditional vs. Integrated Supply Chain Management (Adapted from R. Srivastava, S. Tasadduq, and L. Fahey, "Marketing Business Processes and Shareholder Value: An Organizationally Embedded View of Marketing Activities and the Discipline of Marketing," *Journal of Marketing*, 63, 1999, pp. 168–179)

integrate its own supply chain with that of its suppliers and customers. Following this evolution, we attempted to clearly define supply chain management, as it is practiced today.

The Supply Chain Council, an organization of companies working together to provide a cross-industry supply chain management framework, offers the following definition.

> Supply Chain: The flow and transformation of raw materials into products from suppliers through production and distribution facilities to the ultimate consumer.

The Supply Chain Council has further defined the supply chain by building the Supply Chain Operations Reference (SCOR model, shown in Figure 8.2). This is only a process model and does not clearly show company boundaries or company-to-company structures but focuses on the basic processes involved in any supply chain. This model breaks the supply chain into the core processes of *Source, Make, Deliver,* and *Plan* components and is further defined by more detailed process models within each component area. This "common language" for supply chains offers the opportunity for cross-functional and cross-company communication

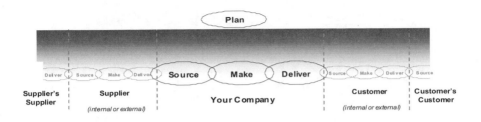

Figure 8.2 SCOR — Supply Chain Operations Reference Model Overview (Source: Supply Chain Council)

and collaboration and is proposed as the preferred supply chain language for the examination of BPO impacts on the supply chain. For this reason, the SCOR model was selected as the framework to be used in examining the impact of BPO on supply chain management performance.

Purpose and Organization of the Study

During the past several years, the concept of SCM has been maturing in terms of theory and practice. Terms such as integrated supply chain management, supply chain optimization, and supply chain collaboration have become the focus and goal of many organizations in the United States and around the world. Global supply chain management has also emerged as a key competitive strategy. Therefore, we posed this question, which guided our research efforts: To what extent is SCM influenced by a business process orientation? The model presented in Figure 8.3 guided our research.

The initial challenge we faced in our study was developing a clear, simple definition of the main concept of SCM. A review of the popular business press literature revealed that SCM was becoming another "buzz word" that seemed to lack a clear simple definition. It is well acknowledged that if you cannot define something in simple terms, you do not know what it is. With that in mind, we developed a definition for this project by first decomposing SCM into its constituent parts:

Supply Chain: The flow and transformation of raw materials into products from suppliers through production and distribution facilities to the ultimate consumer.

Management: The process of developing decisions and taking actions to direct the activities of people within an organization; planning, organizing, staffing, leading, and controlling.

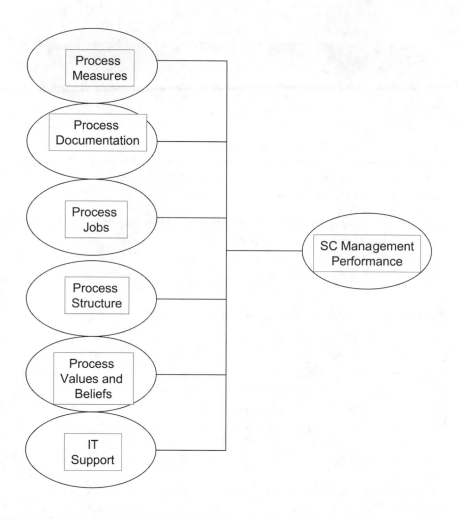

Figure 8.3 BPO Related Relationships to Supply Chain Management Performance

The final definition used in this study combined these two statements to read as follows:

Supply Chain Management: *The process of developing decisions and taking actions to direct the activities of people within the supply chain toward common objectives.*

Defining and Measuring Supply Chain Management

Before we could begin this study, we had to develop detailed definitions and operational measures for the practice of SCM. To accomplish this, we conducted interviews and formed focus groups with supply chain experts and practitioners. Questions were organized generally around the components of BPO but slightly expanded. The questions asked were SCM specific but generally about PV, PJ, process structures, process values and beliefs, PM, IT support, and supply chain specific best practices. The definitions for each of these categories are listed below and are a slightly expanded definition from the original BPO measures discussed in Chapter 5.

Process View (PV): A "Process View" can be described as the process steps, activities, and tasks documented in a visual and written format that creates a cross-functional process vocabulary.

Process Structure: These are cross-functional process team organizational strategies with a "flat" hierarchy, and process owners with leadership, not control-oriented management.

Process Jobs (PJ): These are job strategies that consist of empowered, multidimensional, process team-oriented jobs.

Customer-Focused Process Values and Beliefs: These are organizational values and beliefs displayed through behaviors that are customer-focused empowerment and continuously improvement oriented.

Process Measurement and Management Systems (PM): The components of this area are process measurement systems, rewards for process improvement, outcome measurements, customer-driven measures, and team- and customer-based measures and rewards.

IT Support (IT): This area captures the level of current IT support for the decision processes.

Best Practices: These are activities that are generally presented in the literature, interviews, and focus groups as contributing to supply chain effectiveness or efficiency.

The results of these focus groups and interviews were used to build an initial list of survey questions to be used in this research study.

The initial questionnaire, which was organized according to the 4-SCOR model (see Figure 8.2) processes as well as a section on overall supply chain common themes, was sent to several experts for evaluation and feedback. Wording was modified and some redundancies were eliminated. A revised questionnaire measuring the frequency of the supply chain management activities was then developed using the following 5-point Likert scale (*Note:* the common theme questions used a slightly modified scale):

1 — never or does not exist
2 — sometimes
3 — frequently
4 — mostly
5 — always or definitely exists

The respondents gave their opinions concerning "what, how often, who, and how" activities are conducted their supply chain. Participants were also asked to rate the overall performance of their SCM processes and the performance of each management process by SCOR model category. This initial questionnaire was then tested within a major electronic equipment manufacturer and with several supply chain experts. Wording and format were improved, some items were eliminated, and several were removed based on this pilot test of the questionnaire. A final version of the questionnaire was developed based on feedback from the pilot study and was used to gather the information for our study (see Appendix C).

Data Gathering and Analysis

The sampling frame used in our study was constructed from a membership list of the Supply Chain Council. The "user" or practitioner portion of the list was used as the final selection since this represented members whose firms were in the business of supplying a product, rather than a service, and were thought to be generally representative of supply chain practitioners rather than consultants. This list consisted of 523 key informants representing 90 firms.

The questionnaire was distributed by mail to the supply chain council members with a cover letter explaining the purpose of the study and the sponsorship of the Supply Chain Council. Recipients were also encouraged to distribute the survey to other practitioners within their firms. Forty-three usable surveys were returned. Figure 8.4 shows the industry representation in these responses. Figure 8.5 shows the functional representation and Figure 8.6 shows the respondents' positions within the company.

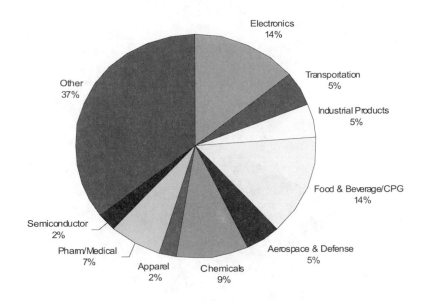

Figure 8.4 Industries Responding to the Questionnaire

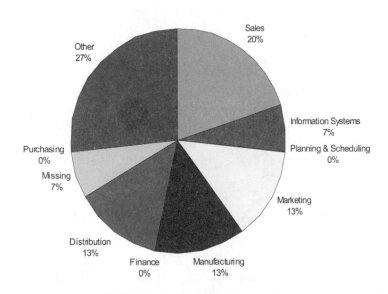

Figure 8.5 Functions Responding to the Questionnaire

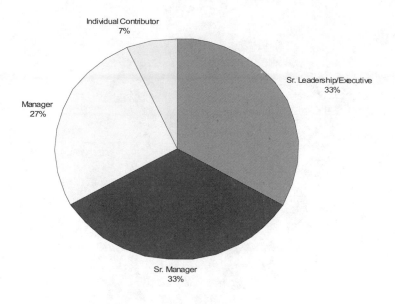

Figure 8.6 Positions Responding to the Questionnaire

As the graphs show, the data represent a broad cross section of industries, functions, and positions. As expected, because SCM is usually staffed with individuals of manager level or higher, the individual contributor portion of the respondents is very low. The function "other" accounts for 27% of the respondents, which represented process-oriented titles for jobs that were cross-functional in nature. This indicates that process-oriented job rating should be fairly high.

The data from the returned questionnaires were tabulated and analyzed. Descriptive statistics (frequencies, means, and standard deviations) were computed, as well as correlation coefficients measuring the relationship of each BPO category (e.g., process structure, process documentation, etc.) to each core process (e.g., plan, source, make, deliver).

Results of BPO Impact on Supply Chain Management

First, general or common theme questions were asked in order to determine the general overall BPO levels within supply chain management. Figure 8.7 shows the results of the five common theme questions.

The results indicate that, in general, the organizations surveyed were fairly evenly distributed across the scale of process orientation, where their mean responses averaged approximately 3 ("somewhat"). Nine percent of the respondents indicated that their jobs are broad process based.

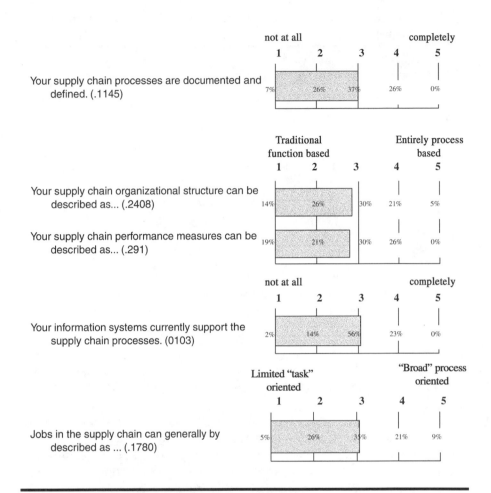

Figure 8.7 Levels of BPO Components in Supply Chain Management

Five percent said that their organizational structure is entirely process based. No one described measures as entirely process based. The level of process documentation or process view also clusters in the middle with no one answering that the processes are completely documented.

The numbers shown in parentheses in Figure 8.7 are the correlation coefficients of each BPO level to overall performance. Process measures had the greatest impact on performance (0.2910), followed by process oriented organizational structure (0.2408), then process jobs (0.1780), and finally process documentation or process view (0.1145). Information systems support does not appear to be a factor. This corresponds generally to our original research but with a few surprises. Process structure appears to be more important than in our original research, while the order of

Table 8.1 Correlations of BPO Components and Core Business Process Performance

Category	Plan	Source	Make	Deliver
Process structure	0.7	0.6	0.5	0.6
Process documentation	0.6	0.7	0.5	0.5
Process values/ Beliefs	0.6	0.5	0.6	<0.5
Process Jobs	0.5	0.5	0.6	<0.5
Process measures	0.5	0.7	<0.5	0.6
IT support	<0.5	<0.5	<0.5	0.7

Note: The stronger the relationship, the closer the correlation coefficient is to 1.0.

the other factors remain consistent. The conclusion from just looking at these common theme questions is that the components of BPO appear to have an influence on SCM. A more detailed examination follows in the next section.

Next, in order to determine the specific impact of BPO on SCM, correlations were performed on the data. Responses to the specific survey questions summed by BPO component category were then correlated with overall SCM core processes. Table 8.1 reports correlation coefficients indicating the strength of the relationship between core process performance and each BPO component.

Examining the detailed relationships reveals a "mixed picture." For classification purposes we drew a line between strong and weak relationships. Above 0.5 are considered strong relationships and below 0.5 are considered weak relationships. Most of the correlations are 0.5 or above while the common theme question correlations were only 0.2. Apparently, when asked in the context of a specific SCOR area, the answers improved in granularity. Asking a question in a more specific context is known to improve the quality of the answer, and in this case it apparently did. Correlations for all components other than IT support were 0.5 or above in most areas.

Process structure appears to be slightly stronger than the others. When we asked respondents about this, we discovered that this was indeed true. The structure represents the span of involvement, influence, and authority in an organization. It is the base operating system for an organization. Like a computer, if the structure does not allow for multidimensional, cross-functional authority then it is difficult to operate. This is particularly true in a management function that demands cross-functional action such as SCM. The basic process structure measures represented cross-functional teaming, process integration, and cross-functional authority of the teams. This makes sense. If SCM is to be successful, the individuals involved must work as a tightly integrated group with shared authority to both make decisions and take action.

Process documentation, according to our research, is also very strongly related to SCM performance (0.5 to 0.7). This is slightly stronger that our original BPO research. One possible explanation is that, in a cross-functional and possibly cross-company activity such as SCM, the documentation of the process to be used is much more important than in other activities. A clear understanding of and agreement about what is to be done seems to be very important in SCM. This is usually achieved through process design and mapping sessions or review and validation sessions with the team. This is a clear message to those implementing SCM strategies. The time and money invested in designing and documenting the processes to be used are critical to success. Omitting this step or allowing it to be done in an ad hoc way will negatively impact supply chain performance.

Process values and beliefs that were measured are actions representing customer trust, firm credibility, and inter-firm collaboration. These appear to also be strongly related to SCM performance (0.5 to 0.6). The *Deliver* area, although slightly below 0.5, is still important. Trusting customers enough to team with them and supply critical information is a very important factor in cross-company collaboration. Trust applies in a similar fashion when dealing with suppliers. For example, it is important that functional employees in an organization jointly participate on operations teams with their counterparts from the supplier firms. Our research also shows that believing what you are told and acting upon it is also a critical factor in SCM. Why bother getting forecasts from your customers if you do not believe them or do not act upon them?

Process jobs (PJ) reflect the assignment of broad process ownership. In this research we measured whether process owners were identified for each SCOR area of Plan, Source, Make, and Deliver, as well as an owner for the overall supply chain. The correlation results of 0.5 to 0.6 indicate a strong relationship between PJ and SCM performance. Clearly, creating broad, cross-functional jobs with real overall supply chain authority is a key component of SCM performance.

Process measures (PM) are also strongly correlated to SCM performance (0.5 to 0.7). This study identified key measures in each SCOR area and asked respondents about the frequency of use. Measures such as supplier performance according to agreements, inventory measures, and customer and product profitability were included in this study. The results clearly show that measures are very important in SCM just as in the original BPO research.

Since many software firms and consultants are emphasizing the importance of IT in SCM, we considered the role it plays in SCM performance. Our research shows that *IT support*, although strongly related to delivery process performance, is only marginally related to overall SCM performance (<0.5). The strong relationship of *IT support* to the *Deliver* SCM process is perhaps because customer order processing and inventory management are usually part of the *Deliver* processes. These are very information intensive processes and, by definition, very dependent on *IT support*. From our research we have concluded that IT investments, by themselves, will not improve SCM performance, except in the *Deliver* process area. Therefore, in order to realize a significant return, these investments must be in support of actions to improve the BPO of an SCM organization.

CONCLUSION

This study investigated the fundamental concepts of BPO in the context of SCM, one of the most critical processes within a firm. The question we posed earlier — "To what extent is supply chain management influenced by a business process orientation?" — has been answered. We also investigated the general level of BPO in SCM. Our conclusion is that, in general, SCM is somewhat process oriented with pockets of excellence. Overall, using the common themes results, process-oriented measures, and structure appear to be strongly related to documentation, values, and beliefs, and IT support. However, when we examined the results by the SCOR processes of Plan, Source, Make, and Deliver, these relationships become more important.

Building the process view and the understanding that results from the construction process is a critical foundation for successful SCM. With the new networked e-corporation it becomes even more important. Establishing the PV across a networked e-supply chain becomes even more important to performance. Putting the process-oriented structures and jobs in place has also been shown to be a key contributor to e-supply chain management performance. Coordinating the flow of materials, cash, and information across a networked group of firms in the e-supply chain is the new competitive battleground. SCM performance is the key to winning this battle.

The results of our research also clearly show that PM and process-oriented values and beliefs are critical ingredients of SCM. Cascading measures used to link people's actions to SCM goals are definitely related to performance but seem to be very difficult to implement. A great number of the participants in our research indicated very little progress in this area. As SCM crosses company boundaries, this becomes even more difficult but may yield a potentially greater return. Process-oriented values and beliefs are also difficult to implement. Trust and credibility are built over time and should not be treated as a "project." Trust is also between individuals, not companies, and is established as a result of hundreds of successful interactions between individuals. Creating an environment that enables this to occur is the task of the leaders of companies in the supply chain and a critical success factor in implementing successful SCM.

Overall, our research has shown that BPO is a critical factor in SCM. The more business process oriented an organization's SCM becomes, the better the organization will perform. This is true for the old economy linear supply chain or the new economy networked e-supply chain.

During the past few years a significant number of companies from many different industries have participated in this research and benchmarking. We invite anyone interested in benchmarking his or her company to contact us for the use of this survey and database. Chapter 9 will outline how BPO can be implemented.

NOTE

[1] Supply Chain Council, *www.supply-chain.org*.

9

IMPLEMENTING AND EVALUATING BPO EFFECTIVENESS

How can you identify, quantify, and improve the BPO level in your company? The previous chapters explained the concept of BPO and how to measure the BPO level in an organization. The impacts of BPO on an organization (improved esprit de corps, reduced conflict, improved connectedness, and improved overall business performance) have also been presented. Several detailed case examples have helped show how to use the BPO assessment instrument and maturity model to identify BPO improvement opportunities in both manufacturing and service environments. Now, in this final chapter, we explain commonsense approaches to developing an understanding of your current level of BPO and propose ways to improve this. Barriers and key success factors are also offered.

THE JOURNEY OF BUSINESS PROCESS ORIENTATION

Most of you have experienced the adventure of a family summer vacation. You begin planning your summer odyssey in the winter. You consult maps and tour guides before deciding on your final destination. Finally, the day arrives and you set off with the car fully loaded. Mom made sure there were plenty of snacks and soft drinks for the trip. Not long after starting out, however, you determine that you are lost and pull over and study the map. Shortly after returning to the road, it's time to look at the map again. Tensions are now rising as the driver (most likely Dad) insists he knows the correct route. Mom, meanwhile, prefers that Dad pull over to ask for directions before they get more hopelessly lost, while the kids impatiently ask, "How long before we're there?" This whole process is frustrating, inefficient, and stressful.

The BPO journey is much like a family vacation. Quite simply, becoming process oriented requires careful planning before you arrive at the final destination. Here are some steps common to both endeavors:

- Defining the end goal of this journey
- Building and understanding the process view
- Knowing your starting point
- Mapping the route and keeping on track
- Strong leadership
- Stopping and asking for directions from time to time

DEFINING THE END GOAL

The first step in this journey is to clearly define the end goal or destination. Hopefully, this book has helped you do that. Implementing a BPO in an organization involves clearly describing the destination. All areas of change should be identified and clearly communicated to everyone involved.

For example, new jobs and their requirements need to be properly defined. Workers assigned these new jobs want to know what will be the impact on them. Multidimensional jobs with broad responsibilities that involve frequent problem solving and learning, all key aspects of BPO, can only be performed by employees who have the ability and authority to perform them. For many people in today's organizations, this is a big change. With a functional orientation present in many organizations the approach to job design was often to limit responsibility and focus on a task. Authority typically rests with the boss, not someone who is actually serving the customer. A BPO, on the other hand, assigns authority and responsibility to those employees who are actually serving the customer.

"How will the work change?" is another important question. Process-oriented work is very different from functionally oriented work. Functionally oriented work too often focuses on performing the tasks that please the "boss." Process-oriented work involves pleasing both the internal customers as well as the end customer, a dramatic shift for many organizations. Customers are concerned with process *outcomes*, i.e., getting what they have paid for. Therefore, pleasing the customer, not political "schmoozing" within an organization, is the gateway to success in a BPO organization. Process-oriented work also crosses functional boundaries. That is to say, functional-based authority is often ineffective outside its function. To be effective in a BPO organization, persuasive or expert authority becomes more powerful. Success in a BPO organization is now predicated on what you know, not on whom you know.

Measuring process outcomes and making decisions based upon them is also a big change for most companies. Much of the power in organizations rests in the budget. The ability to spend money, hire people, and reward people is a key resource allocation function. Allocating resources, meas-uring success, and rewarding people based upon process dramatically shifts and refocuses this power. For example, having a budget for the "Order Fulfillment Process" dramatically shifts the investment focus to process and away from the function. Measuring and rewarding people based upon work outside their "function" or bosses' areas of influence also tears at the functional power structure. Control of the company resources translates into process management control as well. This is why this component of BPO is so powerful and so difficult to achieve. Too often, attacks by the functional power brokers cause the "faint of heart" to fall away at this point, and BPO levels stagnate or collapse.

Educating employees in the organization to understand the benefits of BPO is also critical. A rationale for introducing changes in workflow or job responsibilities must be clearly communicated. Our research has shown that most employees value less conflict, improved cross-departmental connectedness, and esprit de corps. Using the BPO instrument can prove fruitful as a discussion tool. It is absolutely critical to involve all employees in this discussion in order to enlist their energy in the change process.

BUILDING THE PROCESS VIEW

After clearly describing the firm's final destination, the next step is to begin to look at the organization in a new way — through a process lens. Building a process view (PV) should be inclusive, not exclusive, involving most, if not all, of an organization's personnel. Those not involved in preparing the actual documentation should nevertheless review and validate the work being performed. Failure to take these measures will short-circuit the process orientation journey. Moreover, accepting excuses such as not having the time or money to involve anyone except a core team member will stop the effort cold.

Many possible methods could be used to build a PV. An optimal approach begins with a high level, customer-focused process map and works from there to build increasing levels of detail. Figure 9.1 provides a generic example of this type of map.

This map is critical in beginning the shift to a PV as well as a customer focus, both key components of BPO. It is also a very useful tool when organizing the construction of detailed PV, process organization structure, assigning process ownership, tracking team progress, building high level measures, and allocating investments.

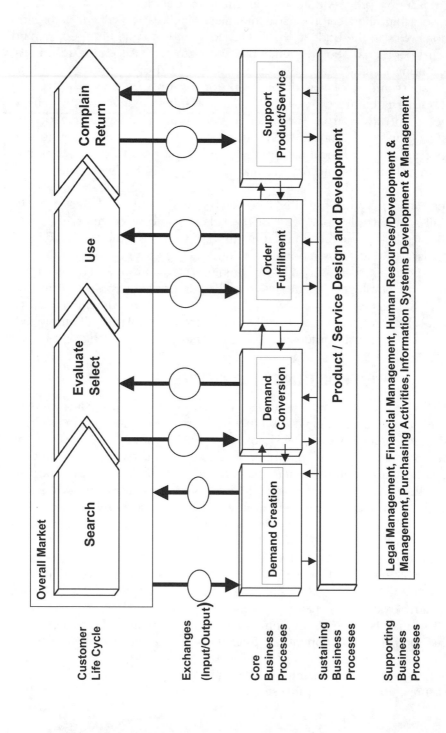

Figure 9.1 A Generic High Level Process Map

Organizing, classifying, and setting boundaries is a very important step in the BPO journey and building this map can help. Organizing and classifying the processes by *customer lifecycle*, the series of activities a customer performs when interacting with a business, helps focus on the outcomes that are critical to the customer and the businesses processes that produce these outcomes. *Core processes*, the value-added activities that support and facilitate the customer life cycle, represent the foundation of most businesses, the value that customers pay for, and the essence of most businesses. The customer *exchanges* or interactions are the inputs that begin the process and the outputs that end the cycle. Measures built around these interactions are, from the customer's perspective, the essence of the business process performance. "What gets measured gets done" is a well-worn saying in process circles. Measures focusing on customer interactions taken from the customer's point of view will more likely lead to real change and superior customer value.

Sustaining processes may not result in direct customer interactions, yet are critical to the operation of the business, such as product research and development. For a chemical company, a sustaining process might be process design; for a consulting company, it may be knowledge management.

Supporting processes are not as insignificant as their position on the map might imply; they are shown at the bottom of the map because they are the furthest from the customer. Human resource management would be an example of a supporting process. In a consulting company they are many times a sustaining process because people are the product to be developed. Information systems frequently serve as a supporting process for many companies today. For example, for a pure Internet company such as Amazon.com, the total customer experience is to a great extent shaped by its Website and its daily function. In this case, design and support of that Website could be viewed as sustaining processes or even as part of a core process.

Using the map in Figure 9.1, process measures can be developed and process owners accountable for these measures can be identified. At this level, each high level process listed on the map should have an owner who is responsible for building the next step in PV. This owner should also take operational responsibility and be given authority for the assigned process. These owners should be on the executive leadership team and involved in key leadership activities, just as the functional executives are. Often, the functional executives become the owners, especially of the sustaining and supporting processes. This owner must then begin to build the process-oriented jobs and cross-functional process teams and lead the transition to them, a tough but not impossible assignment.

Finally, this map also serves as a personality profile of the company, the markets it serves, and the interactions with its customers. It is a critical starting point for building a PV. The next step in constructing this view is to build process maps of increasing detail with cross-functional teams, continually involving more and more of the organization until everyone understands and accepts the processes. It is important to remember that the documentation is not the outcome of this effort; it is the understanding and acceptance by the entire organization. The process owners lead this with the deep involvement of all members of the process team. The result of this effort is then shared with the entire company for validation, understanding, and acceptance.

Building the PV of an organization is very time consuming and expensive but there are no short cuts in this process. Like building a house without a good foundation, an organization without a fundamental PV has nothing upon which to build and will eventually revert back to a predominantly functional, vertical organization.

STAYING ON TRACK

Knowing where you are in your journey and how far you have to go to your destination are very important questions on the route to becoming business process oriented. Using the BPO assessment tool, maturity model, and benchmarking database to answer these questions can be very helpful. At the very least, these tools can help frame the discussion within the organization and these discussions can result in developing the answers.

The steps involved in this effort are:

1. Gathering the initial data and plotting the results on the high-level maturity model.
2. Examining individual BPO component/outcome scores using the BPO maturity model.
3. Comparing the detailed answers to the benchmarking database.
4. Building consensus around actions needed to move forward.

Gathering the data on your specific organization can often prove to be a difficult task. Obtaining cooperation in completing the BPO questionnaire is a major challenge. Also, data must be collected in a comprehensive, yet resourceful manner. Since you cannot ask everybody, asking the right people is very important to the quality of the answer.

This is accomplished by selecting 20 to 30 "key informants" within an organization who can complete the BPO questionnaire (see Appendix C for the actual survey). A key informant is one who knows the answers and whose answers will likely be representative of the organization as a whole. Every organization — each level, function, or geography — has people who have their fingers on the pulse of the organization. They are not necessarily the official leaders but are often the second in line or maybe a continuity leader, someone who has been around a long time and who has had many jobs within the organization. These individuals serve as ideal participants in the survey since they know a lot about the organization, the people, and the processes.

Key informants should be identified from all levels of the organization based on whatever functions or geographies are involved in the BPO effort. Historically, a response rate of 20 to 30% can be expected, so a minimum of 60 to 100 key informants must be included in the sample. More data are not always better, but more responses will improve the representativeness of your sample.

Once completed questionnaires are received from the key respondents, their responses need to be aggregated and averaged. A BPO score is then plotted on the high-level maturity model and detailed benchmarking charts shown in Chapter 5.

Collecting the data is the easy part. Yet, how the results are treated is the key to staying on track. Recently, we went through this process with a major chemical company. After the survey was completed and we plotted the results, meetings were set up with the leadership team. The results were discussed in great detail. This was the "inner voice" of the organization speaking to the leadership on what areas needed more work.

The results were organized by different groups in order to focus in on where efforts were needed and who had to lead them. For example, if PV was an issue in one area then the process owner would be asked to assess the state of this component and develop an action plan for improving it. Decisions were then made based upon criticality of the change and its potential impact. This allowed the leadership team to have an organized dialogue and made fact-based decisions concerning its efforts to improve BPO.

After the leadership team reviewed the results and established action plans, the process owners took the results and recommended action plans to the process teams. The results were discussed and more detailed plans constructed. Specific aspects of the survey were examined in detail. If PV was an issue then the process documents were actually examined. The process teams were asked to validate and present the "process" to the process owner. This helped in two ways. If the documents were flawed then this was a chance to correct them. If employees did not really understand the process, then this activity gave them a chance to learn more about it. Process measures and jobs were examined in the same way.

The use of the fact-based assessment provided a vehicle for examining the progress and issues in the journey to improve BPO. Looking at the results at different levels of the organization can help align the people toward common goals and specific actions.

COURAGE TO CHANGE

As mentioned earlier, the journey to a more business process-oriented organization can be a source of great stress. Change is difficult. The laws of physics say that an object at rest will remain at rest unless impacted with a force greater that the force keeping it at rest. Organizations are no different. It takes a lot of work for any organization to operate from day to day. The force to keep the status quo in place is pretty strong. What can change this?

Without digressing into a discussion of individuals and their motivations, we can answer this question simply by saying most people will do what is asked or expected of them, as long it is reasonable. But someone must ask. This is a key responsibility of leadership, to clearly describe what is expected and "ask" each person to do it. Sorting out the inevitable conflicts between functional and process authority and the struggle to bring them into balance must be ongoing during the BPO journey.

Of course, this is an over-simplification. As mentioned earlier, each employee must understand the goal and path to this goal as well as why it is important. What is the overwhelming reason to change? Employees also must be given the opportunity to voice their concerns and questions and have them answered in a serious way. This is most effectively accomplished through regular one-on-one interactions between individuals and the leadership team.

This is a very difficult process for leaders in an organization to be able to face employees and address their questions, concerns, and fears, and to supply reasonable answers. Business leadership has lost a great deal of trust and credibility with rank and file during the last decade due to downsizing, mergers, and restructuring. It takes courage to stand up in front of the survivors of one of the latest management fads of the past 10 years and say, "Here we go again and it is different this time because...."

Our research shows a clear link between BPO and its organizational impact. Thus, our recommendations are data based, not simply management rules of thumb or "programs of the month." Most leaders would agree that esprit de corps is a good thing and a desirable goal. They would also agree that reducing cross-functional conflict and improving connectedness would be worthwhile goals. Given that BPO leads to a more positive organizational culture and improved overall business performance, it is obvious why company leadership would be interested in embracing business process orientation.

CONCLUSION: BPO AND THE NEW ECONOMY

The Internet and digital technologies have created new opportunities in the new economy. Throughout this book we have endeavored to profile the role of BPO in the new economy.

Our research has shown that BPO is critical in reducing conflict and encouraging greater connectedness in an organization, while improving business performance. In this new economy, competition has shifted to networks of companies cooperating across boundaries in order to achieve market goals. In the past, this was possible only through vertical integration and only in certain markets (chemical, steel, etc.). Now, with the new technologies, this integration is possible in almost every segment of our economy. It is obvious that conflict and connectedness play big roles in this process. Since cooperation is not the norm outside company boundaries (some may say even inside company boundaries), BPO helps engender dialog and "connect the dots" in the network, leading to a competitive advantage. This was a point we made earlier in the book, that companies can match individual processes, but they cannot match the integration or "fit" of these processes among the network players.

Building a common process view between the companies within the new economy network is as critical as the cross-functional view and probably much more difficult. Gaining agreement on process terms, process activities, and outcomes is critical to process integration within and between companies. Building a high-level network process map can help clarify these process roles and responsibilities.

Allocating resources based upon process, a key component of BPO, is also critical to the new economy networks. Companies within the network have to invest in the cross-company processes in order to make the new networked business model work. Decisions concerning who owns the process and the investment necessary to support the process are key to moving forward in the BPO journey.

"Employees must be regarded as assets" is an expression that we often hear. In the networked business, where process-oriented jobs will likely span several companies, making connections among the different network players is the responsibility of individual employees. Employee compensation becomes more complicated in this scenario as well, since it is no longer clear who "owns" the employees and who is responsible for paying them.

In order to move forward in building a new economy business the network must first commit to becoming business process oriented *across the network*. This commitment is critical since it will guide the hundreds of decisions about jobs, investments, and ownership. Some networks are joining together to form separate businesses, called exchanges, in order to house the investments and people supporting the cross-network processes. Independent investors are organizing independent exchanges that exist in the processes between companies in a market. These represent exciting new business models, potentially evolving to pure BPO organizations.

Inspired by the challenges in the new economy and the research, case studies, and conclusions presented in this book, hopefully you have concluded that practicing a business process orientation can result in a key competitive advantage. We also hope that you have gained some insight into why and how new levels of BPO can be achieved. The future competitive landscape is shifting from between companies themselves to between networks of companies. Understanding and mastering process design and change will be tantamount to achieving and sustaining a competitive advantage. Our goal in writing this book was to help prepare you for what lies ahead.

Appendix A

CASE STUDIES

CASE STUDY

ABIG: BUILDING A CORE PROCESS FOUNDATION*

BUSINESS NEED

American Bankers Insurance Group, Inc. (ABIG),[1] a wholesaler of insurance products and services, extended service contracts, and membership programs, is the leading third-party marketing company in the United States[2] and well known in the financial services industry as a product innovator. It has over 200 marketed products available to its multiple distribution channels and national and international subsidiaries. The company enjoyed increasing revenue and profitability growth for many years. One of the contributors to its success is the sales and marketing culture, which originated in the 1980s with the development of a three-stage sales call system.

In 1994, increased growth and product complexities drove ABIG to evaluate its core business processes and the use of new technology to reduce cycle time, improve the quality of output, improve client satisfaction, and create a "knowledge worker" culture. During 1995, a consulting firm was hired to recommend the areas where business process reengineering (BPR) would have the greatest impact on the organization. Several BPR projects were initiated in the sales, and marketing operations, focusing on four of its major cross-functional business activities: sales, contracting,

* This case was prepared by Nancy Rauseo, Ph.D. student, Nova Southeastern University, Huizenga Graduate School of Business and Entrepreneurship.

implementing new programs, and managing the client. The BPR project was introduced in a series of phases, beginning with the preparation stage.

PREPARATION

One of the first challenges for ABIG's top management was to prioritize the BPR initiatives. The voice of the customer served as the foundation for establishing which specific business processes would be first for analysis. A survey was mailed to the top executive of each of the 50 highest revenue volume clients. Each top executive was asked to consult with his/her internal areas for a more accurate view of their expectations and perceptions of ABIG in regard to all their business activities.

After receiving a written response from each client's top executive, an ABIG executive followed up with a visit to each client location. This process allowed ABIG to better understand the issues and demonstrate to the clients that the company was serious about listening to their needs and addressing their issues. The issues of ABIG's clients became the basis for BPR objectives:

- An informative sales process, where clients' problems were clearly defined and addressed;
- Courteous and professional service at every point of contact with the client;
- A single point of contact for all service-related and planning issues;
- Hassle-free installations of new programs;
- Minimum effort on negotiating contracts with clients;
- Greater accuracy in the content of contractual arrangements; and
- A "knowledge worker" culture.

A project team was organized with representatives from sales, marketing, operations, legal, and administration. Three full-time sub-teams were developed to address the core processes of sales, contracting, and client set-up. About three-quarters of the sub-team members were employees with full-time functional jobs. They were temporarily removed from their normal jobs and assigned to the sub-team for periods of time throughout the BPR phases. The other part of the sub-team, made up of members from the administration department, served as internal consultants and facilitators.

With the aid of outside consultants, sub-teams were assigned leaders. Project plans and timelines were developed for each sub-team. The administration department was responsible for overseeing all BPR projects and communicating the progress to executive sponsors. Sub-team leaders submitted weekly reports to the administration area outlining accomplish-

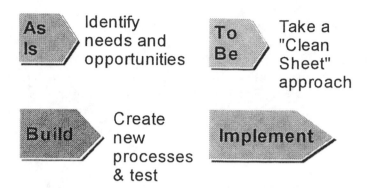

Figure A.1 BPR Stages

ments, issues, and milestones. With the assistance of the consulting company, the three sub-teams progressed through the remaining BPR stages.

AS–IS Stage

During this stage, each sub-team met with various stakeholders and participants of the core process. Through various interviews and meetings with representative stakeholders, the sub-team members collected standard operating procedures, documented process flows, reviewed departmental training programs, obtained organizational charts, and determined resource allocations. A thorough analysis of the collected information allowed the sub-team to develop preliminary, high-level, cross-functional process flows for each core process.

Validation of the high-level process flows was a critical step in this phase. Meetings were held with representatives from sales, marketing, and other areas participating in the core processes. As feedback was received, the sub-team modified the process flows and obtained more detailed information on the lower level processes. This iterative process took approximately 6 months, with the end result documented as cross-functional AS–IS process flows and current resource allocations for each core process. The sub-team also developed an extensive "wish list" from the stakeholders interviewed through this phase. The results of this phase were measured by the following criteria for each of the three core processes:

1. Hand-offs
2. Queues
3. Cycle Time (in calendar months)

TO–BE Stage

All AS–IS perspectives were put aside to develop a vision for the company of the three core processes under study: sales, contracting, and client set-up. The sub-team used the feedback received from the AS–IS meetings and the survey responses from the clients to develop a draft vision statement. Through an iterative process with various stakeholders and executive management, the final vision was completed in approximately 3 weeks:

> *Vision*
>
> *Analyze a prospect's needs, present a strategic and profitable marketing plan, and install needed products and services completely, accurately, and on time.*
>
> *Use technology to create a seamless flow of information between our customers and our people. This will be accomplished through open communications and a networked environment. Data will be captured and stored electronically, decision-making will be computer aided and client data will be stored in our online corporate database.*
>
> *Provide personalized account representation to exceed our customers' needs and expectations. We will have the right people do the right activities at the right time.*

With the vision in mind, the sub-teams proceeded to design the new TO–BE processes. Functional members of the sub-teams were replaced with new ones, to maintain objectivity and new perspectives. Creativity, open-mindedness, and analytical ability were among some of the traits sought when selecting new members for the sub-team. At times, several negotiations took place with the candidates' management to allow full-time participation for this phase.

Another 6 months were spent designing new and innovative processes. Systems analysts aided in the design of new technology that would facilitate the new processes. Creativity exercises were used to encourage a continuous flow of new ideas. Known organizational culture constraints were not considerations during the development of the TO–BE process to avoid limiting the potential of the new process designs.

Again, validation meetings were held with many stakeholders, including SBU management and executive management, since their buy-in and support were critical for the future implementation of the TO–BE processes. In some cases, modifications were made to incorporate new

Table A.1 Percentage Reduction Goals

Core Process	HANDS-OFF	QUEUES	CYCLE TIME (months)
Sales	50%	30%	23%
Contracting	72%	77%	41%
Client Set-Up	43%	21%	37%

business trends and customer needs. The process improvement goals in terms of percentage reduction for the TO–BE process were established (see Table A.1).

BUILD STAGE

Executive management determined that the BUILD phase required ownership by a group or department representing the major stakeholders of the TO–BE processes. The management of the affected functional areas would need to be responsible for selling and encouraging the new processes and all the corresponding organizational and technical changes supporting these new processes. The sales and marketing support area was assigned the responsibility for facilitating the development, testing and implementation of all the changes required to achieve the new BPR goals.

The first step for the sales and marketing support area was to develop a change management strategy to educate employees on the upcoming changes. Both individual departmental meetings and rallies helped to communicate the new vision and the highlights of the changes to come. The executive sponsor kicked off and facilitated many of the initial meetings, making visible the strong management support behind the projects.

It was evident to all involved that the foundation for the BUILD phase was a clear definition of the TO–BE process which was driven and constantly updated based on changing business needs, as illustrated in Figure A.2.

As a result, a nomenclature was created to enable the identification of the organizational components needed to facilitate each of the newly named TO–BE core business processes: winning new business, contracting, implementing new programs and managing the client. Due to the large scope and multitude of projects, the nomenclature was used as a vehicle for effective project planning and management.

The first part of the nomenclature consisted of the various "stages" in each core business process. Each stage of each core business process was

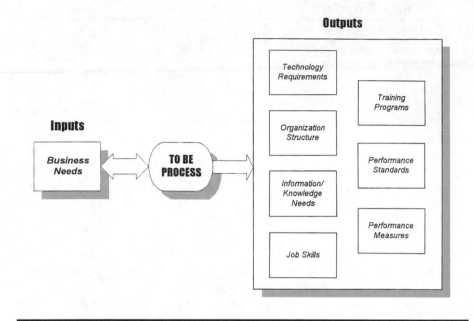

Figure A.2 Inputs and Outputs of TO–BE Process

further expanded into two more hierarchical levels: activities and tasks (see Figure A.3). The technology requirements, organizational structure changes, performance programs, training, service and marketing standards, and performance measurements were all defined at the activity level of each stage in the core business process. The core business process-stage activity nomenclature became the basis for classifying and managing all systems and organizational development projects.

During this BUILD stage, pilot systems and programs were tested in selected areas to ensure that the expected results would be achieved, aiding in the cost/benefit analyses for new systems development and training. The handoffs, queues, and cycle times were continuously measured and benchmarked against the AS–IS performance indicators. Executive management sponsored various incentives to encourage progress toward the reduction percentages expected with the handoffs, queues, and cycle times of each core process, as outlined in Table A.1.

IMPLEMENTATION STAGE

Although the company continues to undergo implementation of changes, the implementation stage began with a core business process program. This program was designed for training all new and existing employees

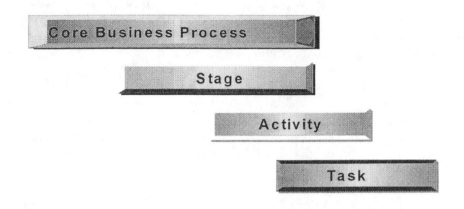

Figure A.3 Core Business Process Nomenclature

throughout the company. A manual documents the core business processes in a user-friendly format. Detailed, complicated process flows were translated into easy-to-understand visual diagrams. The manual also serves as an on-the-job training guide.

The manual was developed by first identifying ABIG's business needs during the BPR initiatives. These business needs were used to create the TO–BE process, which meets the needs of today, and those forecasted for tomorrow. With the identification of the process, the technology needed to support the process was identified as well as the organizational structure necessary to support regular business operations.

This structure, combined with the process, has helped define job descriptions and in turn, training needs. Once the positions and training needs were defined, it became evident that marketing standards needed to be modified to support the corporate vision of how business should be conducted. This manual and its nomenclature have become common language and foundation for communications of all continuous process improvements.

RESULTS

Although there are many BPR implementation projects still under way, the outcomes of this initiative and process foundation have been quite positive. For example, a sales automation system including laptops in the field as well as changes in the organizational structure of sales has resulted in a reduction of 40% in the number of handoffs (almost the goal of 50%). Clients already experience increased responsiveness and accuracy during the proposal process. Proposal turnaround time has gone from approximately 10 days to 3 to 5 days.

Technology has driven many of the productivity improvements being enjoyed by ABIG. The use of Lotus Notes, intranets and common databases have allowed for the sharing and transferring of information and knowledge across functional areas and geographical distances. Online product libraries provide sales executives, home office staff, and multiple subsidiaries with easy access to current, up-to-date product information.

Cross-functional teams have also personalized the customer service functions to clients. Dedicated resources are more knowledgeable about a client's specific business needs. Strategic business plans and service managers have enriched client relationships, resulting in increased satisfaction and profits.

Lessons Learned

- Process analysis comes first. Changes to the organization's structure should follow a clearly outlined cross-functional process.
- Process improvements need to be implemented and tested manually before time is invested in systems development/automation.
- Well-defined processes drive performance measures, job descriptions, and training.
- The process owners must drive the change.
- The process owners must be willing to think outside the box.
- Project management techniques make or break the success.
- Technology developers must have clear specifications — these come from business process requirements.
- A true team approach facilitates the effectiveness of cross-functional processes.
- Executive management support must be visible to all levels of the organization.
- Management must display patience in waiting for the results of BPR initiatives — they take time.

QUESTIONS

1. What were the most significant factors affecting the results of the BPR projects?
2. Why is it important to clearly outline the process before beginning system development work?
3. How did the make-up of the BPR teams affect the projects?
4. What impact did client feedback at the onset of the BPR initiative have on the results?

5. How can a BPR initiative, such as the one followed by ABIG, lead to sustainable competitive advantage?
6. What should ABIG do to ensure that results continue in the positive direction?

NOTES

[1] American Bankers Insurance Group, Inc. (ABIG) was acquired by Fortis, Inc. in August 1999. Fortis merged ABIG with the operations of Fortis-owned American Security Group, forming Assurant Group. Assurant's member companies have more than 110 years of combined experience in the specialty insurance industry, operating in the United States, Canada, Latin America, the Caribbean, Ireland, and the United Kingdom.

[2] Source: *The 36 Leading Players in Insurance Mail-order Marketing*, compiled from a 1996 study by Donald R. Jackson, "Insurance Direct Marketing: 1996 Special Report on The Companies, The Practices, The Standards and The Benchmarks," *National Underwriter*, Nov. 4, 1996.

CASE STUDY

BOSTON MARKET*

When the clock struck twelve noon, the employees in the JW office building let out a holler. "Lunchtime!" they shouted. "The highlight of the workday." With only half an hour for lunch, their dining choices were limited. Many employees "brown bagged" it, but just as many wanted to go out for lunch. Fortunately for the workers of the JW office building, there was a string of fast food restaurants across the street.

One of the restaurants was Boston Market. A group of office workers decided to try out the place. They offered to bring back lunch to several of their colleagues. They had been to the McDonald's next door on numerous occasions and had plenty of time to eat lunch and get back to work. The same should hold true for Boston Market. After all, Boston Market considered itself to be a quick service restaurant. This process shouldn't take long. Or should it?

At 12:05 the group arrived at Boston Market. When they got in the line, there were six people ahead of them, not uncommon for that time of the day. Almost 5 minutes elapsed before Sarah, the first member of the group, placed her order. It was now 12:10. After placing her order, Sarah shuffled her way through the "L" shaped line. Since all food is made to order, it would take a few minutes to prepare her sandwich. A separate employee took her side dish order while yet another employee served as cashier. Just before she got to the cashier, still another employee grabbed Sarah's completed order from the sandwich window and placed it on her tray. Sarah couldn't help but notice how all the employees working behind

* This case was prepared by Alan Seidman, Assistant Professor of Hospitality at Johnson & Wales University, North Miami, Florida.

the line kept bumping into each other. The process seemed disjointed and the line moved very slowly.

Sarah finally reached the cashier where she paid her bill and received a paper cup for her drink. She would have to fill her drink order on her own. Four minutes had gone by from the time she first placed her order until the time she received it. By the time she got her drink and found a table, it was 12:14. Almost half of her lunch period had expired.

One by one, her colleagues worked their way through the line and sat with her. It was almost 12:20 before Kenny, the sixth member of the group, sat down. This left them 5 minutes to enjoy their lunch. What good was that?

Then Sarah remembered that they had neglected to order the food they had promised to bring back to their colleagues. The group turned and looked at the line. It was even longer than when they first arrived. "Forget it," Kenny said. "We'll have no time to get their order and get back to work on time."

This case study examines the operational system employed by Boston Market. More specifically, it looks at the history of the company and concept, the queuing system, the production system, and where the company is today.

Company History

The first Boston Chicken restaurant opened in 1985 in Newton, Massachusetts. The original concept, a fast-paced operation offering home-style foods, was pioneered by Arthur Cores and Stephen Kolow. Cores had experience working in a gourmet grocery store, while Kolow's expertise was in real estate. Their original menu contained marinated chicken, an array of vegetables and side salads, chicken soup, oatmeal cookies, and sweet corn bread. Their small restaurant was an instant success.

In 1989, George Naddaff, a local venture capitalist, met with Cores and Kolow and successfully convinced them to expand their business. In 1990, they had expanded to 13 restaurants with another 15 slated to open in 1991.

By 1991, the Boston Chicken concept caught the attention of Saad Nadhir and Scott Beck, two former executives of Blockbuster Video. A year later, they purchased a controlling interest in the company. Shortly after, the chain's headquarters was moved from Boston to Chicago (and later to Colorado), and a staff consisting primarily of former Blockbuster executives was assembled. They planned an aggressive growth strategy, changing the names of the individual stores from Boston Chicken to Boston Market, reflecting the broader range of new menu items. Corporately, however, they were still known as Boston Chicken.

In November 1993 the company went public and made its Wall Street debut, selling for $10 per share. It closed at over $25 per share that day, raising more than $54 million. In 1996, Boston Chicken stock reached a high of over $40 per share and the company was opening the equivalent of one store every day. The company finished the year with over 1100 stores, bringing in close to $1.2 billion in annual revenue.

The Quick-Service, Home Cooking Concept

Boston Market's concept is unique. It combines an atmosphere of casual dining and home cooking with the convenience of a quick service restaurant. The restaurants offer traditional, home-style products such as chicken, turkey, ham, and meatloaf, which can be served as a sandwich or as part of a platter. They also offer a wide variety of accompanying side dishes, including corn, rice and beans, potatoes, spinach, mixed vegetables, green beans and stuffing. In addition, the menu features chicken pot pies and chicken soup.

Boston Market's biggest success initially was not its restaurants or personnel, but the creation of its own market segment. Almost single-handedly, Boston Market developed the market segment that is now known as "Home Meal Replacement."

Home Meal Replacement, or HMR, has evolved from the changing demographics of today's society. Women have joined men in the workforce in ever-increasing numbers. Gone are the days when the wife stayed home and prepared dinner for the family. Today's consumers do not always want to eat in restaurants nor do they want typical fast food or necessarily prefer to cook. Increasingly, demographic studies point to consumers' desire to cocoon, or stay home. Boston Market, the leader in HMR, was clearly in a very desirable position to capitalize on this trend.

System

Boston Market's service is built around a single channel, multiple phase queue line (see Figure A.4).

As pictured in Figure A.4, customers enter the building and get into a line that forms along the side of the building. Once at the front, the queue takes on an "L" shape. It is along this perpendicular crossing that an available service attendant greets you and takes your order. This is the first subsystem (Subsystem #1). If a sandwich is ordered, the attendant writes up a ticket and calls it back to the cook.

The second substation (Subsystem #2) is where the food is prepared. If the order is a platter or a pot pie, the attendant takes a plate and begins filling it with the side items of your choice. Approximately 10 to 12

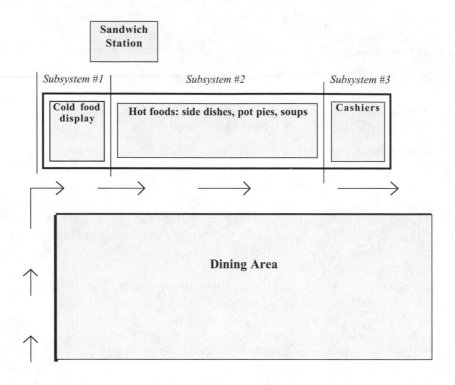

Figure A.4 Boston Market Queue and Service Flow Layout

different side items are available. They are not listed on the menu but instead are displayed in the hot food case about halfway down the service line. The chicken, meatloaf, turkey, and ham are kept sliced and under heat lamps behind the line next to the sandwich preparation area. To fill these items, the server turns around and places the order on each plate. If the order is not ready, the plate is left under the heat lamp until the cook completes it. A sandwich generally takes a bit more time to prepare. Once prepared, sandwiches are presented to the customer further down the service line. During this time, the tray of food always remains on the service side of the window and not in the customer's possession.

Once the plate is ready, the customer moves to the third subsystem (Subsystem #3). This is at the end of the line where payment is made. Boston Market generally has one to three cash registers in operation during peak revenue periods. At this time, the order is finalized. If a drink is ordered, customers are given cups that they fill themselves at a separate beverage area. Coupons are also presented here.

During peak revenue periods approximately five service people and cashiers work the front line. Usually the cashier's position is designated and he or she remains stationary. The other service personnel tend to float around, randomly assisting customers or performing any tasks as needed. Each customer is approached by a different service person to begin the order. Sometimes that person sticks with the customer through-out the duration at the first substation and sometimes the customer is passed on to one or two other employees. There is no continuity during the service flow.

When the cook finishes preparing a sandwich, he places it on a ledge until an employee picks it up and brings it to the customer. There is tremendous variability in the service time for sandwiches. Sometimes they are presented to the customer shortly after ordering. Other times, the customer has already paid and is waiting at the end of the counter for his or her sandwich.

During this process, the customer usually receives some degree of personal attention. Unlike other fast food systems, no number is given to the customer. The service attendants match up each order with each person to the best of their ability. This is in keeping with Boston Market's approach toward maintaining a more upscale position in the quick service industry.

System Challenges

During peak revenue periods (i.e., lunchtime) customers can face long waiting times in the queue. When someone first enters the line, the waiting time for a service employee to greet you and take your order can be anywhere from 45 seconds to 1 minute per person. If a customer is the eighth person in line, he or she can expect a waiting time of 6 to 8 minutes before being approached for service. After placing the order, the customer has to wait again until it is completed. Depending upon the order, the wait can be anywhere from 3 to 5 minutes (on average). This means a total waiting time of 9 to 13 minutes if the customer is the eighth person in the queue. If he or she has 30 minutes for lunch, the waiting time alone takes up a substantial portion of time.

Bottlenecks within the queue line at the various substations present another challenge. Production bottlenecks are obstacles to increased out-put. They can be classified as either episodic or periodic. Episodic bot-tlenecks include machine breakdowns, material shortages, and/or labor shortages. Chronic bottlenecks result in problems inherent in the process, including insufficient capacity, quality problems, poor layout, and/or an inflexible work process. Because of the prevalence of labor shortages and equipment failures facing the quick service industry and failure of the system's capacity to meet demand, Boston Market faces both episodic and chronic bottlenecks.

All quick service restaurants suffer some degree of bottlenecking due to the unpredictable nature of the demand as well as periodic labor shortages. However, Boston Market seems to suffer from a chronic bottlenecking problem. In the Boston Market system, if customers behind in the queue get their orders processed first, they move ahead of customers who are in front of them in the queue line. If there is a question or problem during subsystem one, the entire queue is slowed down until the situation is remedied. Likewise, if there is a problem or question during subsystems two or three, the queue is slowed down and bottlenecking can occur.

Another problem is the lack of sequential consistency. It is not a system of first in, first out. Orders are completed based on the type of order placed, menu availability, prep schedules, etc. Because the service system is unregimented and disjointed, customers often end up "jockeying" among each other. This creates an aura of confusion and inequity. Although it is probably not completely avoidable, a system should be in place that would keep this to a minimum.

Lastly, during peak revenue times, there can be a great deal of confusion among the employees. Most peak time periods feature employees bumping into one another, communicating with one another, and mixing up orders. This is largely due to the disruptive nature of the service flow.

The Unintended Consequences of Coupons

Because Boston Market started out by concentrating on its dinner business, sandwiches were not part of the original menu. They were added in 1994 as an effort to target more lunch customers. At the same time the company began offering discounts in the form of coupons. Coupons offering discounts on lunch and dinner items began appearing regularly in local newspapers and mailers throughout the country. In 1997, the more elaborate Carver Sandwiches were introduced and the flow of coupons became even more aggressive.

Although the coupons were successful in increasing lunch traffic, they created other problems for the company. From a strategic point of view, it changed the position of the company from that of a more upscale fast food alternative to an operation appealing to the masses. Whereas before the coupons, Boston Market used to consider supermarkets and "sit-down" restaurants to be its direct competition, it now aligned itself more closely with McDonald's, Burger King, Taco Bell, and other traditional quick service restaurants.

Another problem brought about by coupons was the logistics involved in collecting them. They were to be collected by the cashier (at the end of the queue) who would apply the discount to the order total. Customers would often forget to present the coupon to the cashier as a result of the confused nature of the queuing system. Customers would often tell the person who took their order about the coupon (who usually expressed little or no interest) and would forget to tell the cashier later on. This presented even more confusion and frustration for both the company and the customer.

In an effort to de-emphasize the lunch business and reemphasize the more profitable dinner trade, the company stopped issuing coupons in 1999. This strategy was short-lived, however, and coupons were reintroduced a year later.

EPILOGUE

In 1998, with its debt escalating and its profits and stock price plummeting, Boston Chicken, Inc. filed with the United States Government for Chapter 11 bankruptcy protection. Shortly thereafter the company closed 178 stores. Boston Chicken stock, once as high as $40 per share, dropped to below 50 cents per share.

The fledgling company was acquired by McDonald's Corporation for $175 million on May 26, 2000. McDonald's intends to keep the Boston Market concept alive although it does plan on converting many of the stores to other concepts.

The financial failure of the Boston Market concept had many causes. Failure to implement and execute a desirable operational system is one cause. Similar businesses should take note of this and learn from Boston Market's failures (see Table A.2 and Figure A.5).

Table A.2 Boston Chicken Inc.: Selected Financial Data 1993–1997

	Net Sales (000s)	Net Income (000s)
1993	42,530	1,647
1994	84,519	16,173
1995	126,228	33,559
1996	199,460	66,958
1997	378,934	–223,892

Net Sales & Net Income

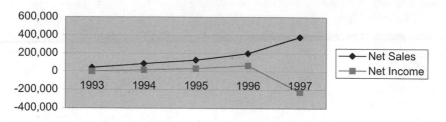

Figure A.5 Boston Market Net Sales and Income

QUESTIONS

1. Compare the queuing system at Boston Market to that of other quick service restaurants. In what ways are they different? In what ways are they similar?
2. How could Boston Market have made its queuing system more equitable and operationally efficient?
3. What were the implications of the aggressive use of coupons on Boston Market's overall strategy? Do you agree with this tactic used by Boston Market? Why or why not?
4. Comment on the importance of defining, measuring, and improving organizational work processes. Approach your response from both the company and the customer perspectives.

CASE STUDY

NEW SOUTH, INC.: CREATING A CONTINUOUS IMPROVEMENT CULTURE*

INTRODUCTION

New South, Inc. is one of the 20 largest (in sales) lumber manufacturing companies in the United States. It is also one of the five largest privately held lumber manufacturers in the country, and, since its inception, it has been managed by members of the founding families. As a producer of Southern Pine products, it is one of the largest and most successful independent manufacturers in the nation. When the company was founded in the mid-1940s, its competitors numbered in the several thousands. Today, there are approximately 500 competitors.

There are four manufacturing facilities in South Carolina and North Carolina. For over 40 years, New South, Inc. has impressed customers and competitors with its high quality lumber products and superior service. Since it started in 1946 as Waccamaw Lumber & Supply Company, the organization has revolutionized the sawmill industry with innovative systems for optimum log utilization. Over the years, it has applied modern technologies to extend the natural resource, timber, to make as much lumber as possible from the timber supply.

---

* This case was prepared by Kitty Preziosi, Corporate Catalyst, Preziosi Partners, Inc.

A key requisite to being successful in this business is to ensure that none of the timber is wasted. Efficient handling of this input requires technology and sound management decision making in this often around-the-clock manufacturing enterprise. The average input required for each lumber manufacturing plant (sawmill) to produce 600,000 board feet (roughly 60,000 pieces of lumber) per day is 8,000 to 10,000 logs.

New South is well positioned to ship its products in a timely and efficient manner to meet customer demand. A fleet of over 50 company-owned trucks and railroad connections gives all plants convenient access to the eastern United States and midwest. It also has a thriving international division, as its products are used in the production of a variety of wood products.

Creating a Continuous Improvement Culture

Key efforts by New South management and employees were made to create more effective and efficient management and production processes, sustained by a continuous improvement culture. Four initiatives were introduced to effect these changes:

- Train and coach managers and employees in its unique molding and millwork plant.
- Train the senior team in creating and managing a continuous improvement environment.
- Employ process improvement teams to analyze and improve key work processes.
- Develop a company vision, including core values, core purpose, a 20-year bold quest, and a vivid description of the quest.

Training and Coaching in the Molding and Millwork Plant

In 1996, the company was struggling with its new molding and millwork plant. This plant was unlike other company plants and required knowledge and skills not already present in the company. The plant was consistently losing money. It was the only plant of its kind in the southern United States and required much attention to get a return on investment as soon as possible. A molding and millwork expert was brought in to assess the problems and make recommendations for improvement. One of the recommendations was to train the plant management and employees to create a continuous improvement environment. The consultant recommended bringing in QualTeam, another consulting firm specializing in continuous improvement.

A QualTeam consultant assessed the situation and made the following recommendations: (1) To train and coach the plant employees and managers in the practices and processes of continuous improvement and (2) to train and coach New South's senior management in their role to lead the development of that continuous improvement culture. New South's senior team agreed to the work plan for a 6-month period and would assess progress and value at the end of that 6 months.

QualTeam began working together with the plant personnel to determine the goals of the effort. Everyone in the plant was aware of the string of successive monthly losses and wanted to play a role in turning around the situation. The consultant, management and employees put much effort into determining what should be measured in the various processes, how they would measure it, and how they would determine levels of success each day. Each work center had production goals and yield goals. Within 4 months, significant improvements were achieved, including a 21% decrease in waste and a 60% decrease in rework caused by errors in marking the input for cutting.

While the production and yield results were significant and employee morale continued to improve, senior management realized that process improvement alone was not sufficient to meet the new, price-driven foreign competition. The company president determined that in a "reasonable market," these continuous improvement efforts at the plant would have worked. However, the internal operations of the plant were deemed not to be the fundamental problem. The bottom line was that the molding and millwork plant could not make money, given aggressive foreign competition. Therefore, the plant's purpose was changed to focus on other products that could add to the profits of the company.

Senior Team Training and Coaching

The work with the senior team, however, proved to be a good investment of time and money for the organization in the short and long term. The president, who is the son of the company's co-founder, said that he wanted his team members to appreciate the need for continually improving their skills, to look at things differently than they had looked at them in the past, and to see change in a positive light. The president said that although the company had not provided much training in the past, he firmly believed that training and developing people sends a message that you care about them. He and his team asked that skills improvement be focused on the following four areas:

- To work together as a team
- To be more effective communicators

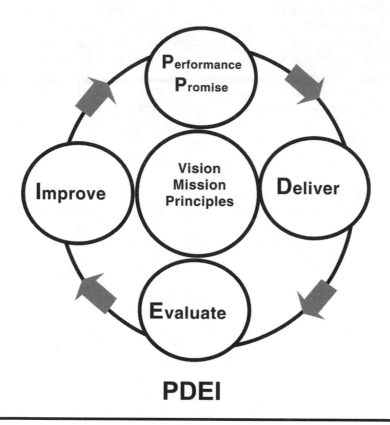

PDEI

Figure A.6 The Continuous Improvement Cycle (Adapted from Robert F. Lynch and Thomas J. Werner, *Continuous Improvement: Teams & Tools*, Atlanta, GA, Qualteam, Inc., 1992)

- To solve problems more effectively and efficiently
- To gain additional appreciation for empowering people.

The consultant trained the senior managers in such topics as PDEI (Performance Promise, Deliver, Evaluate, Improve): the continuous improvement cycle (see Figure A.6), leadership requirement, process management, measurement, team development, communication, team effectiveness, problem solving, fundamentals of statistical process control, team decision making, and self-management. The team members were given homework to prepare for the next session to help ensure transfer of learning to their jobs. QualTeam also trained three selected New South managers in an intensive 1-week course on internal consulting in order to strengthen the organization's ability to develop continuous improvement competencies throughout the 800-employee organization.

Four years after the training of the senior team began, the president reported that the training helped them define the roles of various members of the team. It also helped them develop "winning ways," leading to improved teamwork and improved communication within the team as well as throughout the organization. Of particular note was the increased emphasis on fact-based problem solving vs. intuition-based problem solving used in the past. Since the training began 4 years ago the senior team has continued to have regular meetings with a clear delegation of decision making and accountability. Senior management roles have changed. New people have been brought in with positive attitudes toward change and valuable experience not enjoyed by other members of the team. After 4 years of growing the new culture, the president stated that attitude and ability to adapt to change required to create a continuous improvement environment had indeed been realized.

Analyzing and Improving Key Work Processes

As the senior team studied ways to improve the productivity and quality of life within the company, it became clear that several key work processes were ripe for analysis and improvement. Two process improvement teams were, therefore, created. One was dedicated to studying the financial processes (accounts payable, accounts receivable, general ledger) with the idea that a new general ledger system could be installed to improve those work processes. The other group was dedicated to studying two key processes of the sawmill operations.

The initial focus of the finance process improvement team was learning techniques and tools of process mapping to analyze the three financial processes, isolating problem areas and resolving those problems and, in general, making changes to streamline those processes.

The initial focus of the production process improvement team was on the handling of log receipts and storage at the plants – the first step in the lumber production process. The team was instructed to examine each process well enough to determine if it needed to be improved, to reduce costs where possible through streamlining, and to ensure that all steps in the processes were "value-adding." The team was admonished not to overlook anything with the excuse that "we have always done it that way." Thus, a series of team ground rules were issued.

An important ingredient of kicking off each process team was a team consensus on meeting ground rules. Because each team was made up of selected experts from a variety of disciplines within the company, it was important that everyone be "on the same page" regarding how to operate as a team in the analysis endeavor. Figure A.7 shows an example of the ground rules developed by one of the teams. This list was posted in a prominent spot at each of the meetings.

Ground Rules

- Be open to new ideas. Don't cling to old ways.
- Be considerate of others in the group.
- At the end of each meeting, schedule the next one.
- Ask questions if it's not clear.
- On major points, gain consensus among all.
- There are no stupid questions.
- No meetings the first week of any month.
- Leave the meetings with clearly defined action plans.
- Complete all assignments on time.

Figure A.7 Ground Rules

Process analysis began by evaluating the big picture, to agree on the purpose, boundaries, suppliers, customers, inputs required of suppliers, and customers' requirements. Figure A.8 shows the elements for the process improvement team that focused on the log receiving and storage process.

The next steps were to create a current status map, analyze the current status, identify improvements, map the ideal state, and define measures and management needs. The entire process mapping was done on the wall using large sheets of white paper with various colors of Post-it™ notes to depict the steps of the process.

Often such process analysis initiatives experience resistance when attempted in organizations. There was little resistance in this case primarily because the president and other key senior operating managers were the champions of the initiatives and helped the consultant and team members focus on the right track. These initiatives, in fact, were a part of the objectives of those senior managers. In other words, once the continuous improvement initiative was in full swing, these senior operating managers asked for help in analyzing their key processes to make improvements in efficiency and quality.

Senior Team Creation of a Company Vision

As the senior team members participated in their leadership for continuous improvement training, it became clear to them that they needed to re-visit their strategy to ensure that what they were learning and applying from the training was reflected in their strategy. Upon further conversations about the strategy, the team decided to create a long-term vision for the company plus a set of values they believed in for running the type organization they would be proud of. Thus, an unintended but valued outcome resulted from the senior team training.

Process Improvement Team
Log Receiving and Storage

Purpose of This Process
To inspect, measure, and inventory logs in an efficient manner in order
to provide the sawmills with consistent quality volume and guarantee
timely and accurate payments to our suppliers, consistent with New South
specifications and procurement strategy.

Process Boundaries
Beginning: at the scale house when log truck arrives
Ending: at the log deck

Process Suppliers
Loggers, Dealers, Landowners, Landscapers, NS Procurement Department

Inputs Required of Process Suppliers
Raw material: Tree-length logs and pre-cut logs
From Procurement: Schedule, pricing, inventory strategy

Process Customer
Log Deck

Process Customer's Requirements
Adequate supply as required by the sawmill
Logs placed on deck correctly

Figure A.8 Process Improvement Team Log

The consultant recommended the approach Collins and Porras took in
their book, *Built to Last*. The consultant facilitated the team members to
discover their core ideology (organization values and organization purpose) and define their envisioned future (as articulated in a bold daring
quest). Each of these components is explained more fully below.

Values. Discovering organizational values by:
■ Sharing their individual personal values
■ Developing team values
■ Discussing what philosophies and practices have made New
  South great
■ Writing and explaining the Value Statements that are true now
  and that they wanted to perpetuate

Purpose. Developing the purpose by discussing:
■ Why New South exists
■ How the company has evolved
■ How New South achieves competitive advantage
■ Which motivations continue to contribute to the company's success

Bold Daring Quest. The team came to a consensus on New South's bold daring quest and a vivid description of that quest by:
- Discussing their aspirations for the company in the context of the company purpose
- Imagining the year 2020 in terms of:
    - Markets
    - Environment
    - Customers
    - Regulation
    - Technology
    - Alternatives to Wood
    - Competition

The senior team members then took this vision to several stakeholder groups. First, they presented it to a meeting of middle managers plus members of their board of directors. An electronic version of the delphi method was employed for that meeting of 30 people to achieve thorough and active participation by all participants. Following that feedback meeting, the senior team, as a group, presented the vision to every employee of New South in small group meetings.

RESULTS

Following completion of the team's work, team members prepared a presentation to senior management to report on their work, their findings, and their recommendations. The finance team's work yielded a much broader result than originally anticipated. The team determined that the company needed to move to centralizing and combining duties in order to prepare the company for future growth. Therefore, instead of focusing on improving the financial processes as a stand-alone entity, the team made the case for designing and installing a new enterprise type system to include not only the financial functions, but also to cover the functions of sales, trucking, timber, and international. The team used the data from the process analysis to convince senior management to buy software to assist in the selection of possible vendors to provide an enterprise system.

That system is presently being installed. The $1.5 million cost has been justified because of significant cost savings as well as the already realized ease of training the personnel in a newly acquired plant. The work now being conducted to implement the system involves several cross-functional teams. Reports from the project manager and the president confirm that this across-the-board involvement in the analysis and implementation is money well spent to ensure buy-in and optimal application.

The log receiving team's work resulted in the generation of much more accurate, reliable data for this important step of the manufacturing process. The receiving process became more automated and the company had improved information in terms of evaluating purchases from its 50-plus timber suppliers. Since 70% of the cost of operating a sawmill is in the timber cost, savings in that area had significant impact on the bottom line.

The company continues to value the work of process improvement teams. Since the first team began several years ago, the company has continued to invest the time and resources of its top experts to study problems and make data-driven recommendations to senior management. The president and other members of the senior team are clear in their desire for improved decision making, which will continue to produce increased efficiencies in the plants as well as in the support teams.

When the chairman of the board of New South tragically died a short time after he participated in the development of the vision, the board of directors of New South and other shareholders were eager to know about the plans and sustainability of the company without its respected chairman. The president used the continuous improvement initiatives — the vision, the process team successes, and the senior team management skill building — as the context to discuss the current state and future strategies of the company. He and the senior team were committed to make continuous improvement a vital part of the planning and implementation process at New South during those critical discussions with board members and shareholders.

Looking back on the vision-building work 2 years ago, the president believes the group members did a credible job of evaluating what they do, where they want to go, and what the vision really is. There is also much evidence that they did a credible job of rolling out that message to all of the New South associates to get the thinking ingrained in every employee. The Vision is still an integral part of their planning process — in the budgeting and strategic planning. The president stated, "By putting this on the line we are obliged to make it a reality. Execution is requisite, especially as we grow our business to meet our "2020 Vision."

Overall, the company president knows that the continuous improvement initiatives have enabled the company to run its operations better. He is delighted to report, "We are having more fun working together. It's priceless."

KEYS TO SUCCESS

There are several keys to success for such an initiative that were a reality during these projects at New South, Inc.

- The president, Mack Singleton, provided thoughtful, consistent, and patient support for the purpose and activities of the initiatives. His patience and belief in the people and possibilities helped the projects stay on track, especially in the most difficult times. He was public and frank in his championing of the cause.
- Senior management committed the best internal experts and representatives to serve on the cross-functional process analysis teams.
- The initiatives relied on and trusted the people in the organization to make the action plans realistic yet challenging to achieve the lofty goals. Respect for the people involved is always very important when asking people to try something new.
- The overall initiative included the classic core pieces of organizational change such as the company values and vision articulated from the top down, integration of the continuous improvement principles and practices in the strategy and business goals, frequent public championing of the cause ("walking the talk"), skill development starting at the top, and empowerment of selected internal experts to solve problems. The sequence of these elements did not take the classic route in New South. (For example, the vision and values project started over a year after the first project.) In the end, the order did not matter as much. What mattered was the level of authenticity devoted to each element and that all key elements were ultimately implemented, albeit in a unique order.

QUESTIONS

1. How would you assess New South's commitment to process improvement?
2. What factors in this case led to the company's financial success?
3. Discuss the relationship between corporate vision and process improvement at New South.
4. Why are process improvement teams better suited than an internal corporate process improvement department to address process change and improvement?

CASE STUDY

TIME INSURANCE: A STUDY OF PROCESS QUALITY IMPROVEMENT*

A longtime industry leader in providing health insurance to individuals and small groups, Time Insurance Company had always enjoyed an excellent reputation with its network of independent agents. But increasing costs, added product complexities, and uncertainty in the health care industry threatened Time's ability to maintain its track record of profitable growth. In the second and third quarters of 1992, the company's management undertook a fundamental review of its business strategy, which included a 6-month reengineering effort in the individual medical underwriting unit. Dramatic improvements in quality were achieved through a combination of strategic context, methodology, teamwork, and commitment.

FORMING A TEAM

The company's main challenge was to increase effectiveness in dealing with an increasingly uncertain and changing environment where local and regional differentiation requires rapid and flexible competitive actions. This meant the organization had to become more nimble in identifying

* This case was prepared by John Feather, Partner, Corporate Renaissance, Inc. and Bill Johnson Professor of Marketing, Huizenga Graduate School of Business and Entrepreneurship, Nova Southeastern University.

Figure 1. Phases of the Project

Phase I	Phase II	Phase III	Phase IV
Planning	Analysis	Design	Implementation
0.5 months	2 months	2.5 months	

Figure A.9 Project Phases and Time Lines

and taking marketplace initiatives, while simultaneously achieving substantial improvements in operating costs and service. A critical element involved the identification and redesign of Time's key business processes to simultaneously minimize cycle time and waste while providing superb quality service to policyholders and agents.

It had become quite apparent that Time's medical underwriting department was in trouble. Policy issuance for the last 2 years remained flat, units costs were increasing, and policy reissues had reached an alarming 10%. Moreover, help-desk calls for application-related problems were increasing 50% per year. Incremental improvement would not help; radical change was needed.

A project team was formed with nine Time employees and two consultants. The two consultants acted as reengineering czars, providing project management guidance, while facilitating process analysis. The team was charged with developing and testing a new process design. Figure A.9 illustrates the phases of the project and the approximate time spent on each phase.

Planning the Project

First an organizational readiness assessment was conducted to determine the company's climate for change. The assessment was done by the consulting firm using a questionnaire and interviews. The results indicated that the organization was change-ready, meaning that most employees recognized the need for, and welcomed, change.

Next, biweekly divisional meetings were held to inform and include the entire organization in the project. Monthly communication forums were also established, as were daily mechanisms for employee involvement, including a newsletter, an electronic mailbox for questions and ideas, a suggestion box, and a "living list" that included ideas that employees thought should be incorporated into the design.

Analyzing the Situation

As part of the analysis, the team developed a customer segmentation analysis (needs assessment by customer type), a workload profile (volume and mix of work), activity-value analysis (steps in the process that add value for the customer), design specifications (specific customer demands, such as 24-hour turnaround), and design options (range of options used to customize the design). It then constructed a business process map that detailed work flow to the customer.

From the start, the team felt it needed to look at the new policy issue process from the perspective of its customers: the agents. To gain this outside-in view, it developed a business process map. In this map, the flow of work required to issue a new policy was described in terms of blocks of activity. For example, as applications go through the process, each underwriting request gets "sold, processed, delivered, and serviced."

By mapping work flow, the team uncovered startling information about the underwriting process. First, it learned that contact between the company and the customer was minimal, with lengthy gaps between each intersection. At the time, the process required the underwriter to wait for requirements to arrive, and it took an average of 37 days of internal processing time to issue a policy. Often as much as 60 days had elapsed by the time the policyholder received the policy.

Mapping the existing process also revealed that a new policy application went through 284 process steps, only 16 of which actually added value for the customer. These 16 steps accounted for only 9% of the process time (the actual hands-on time that a person spends working on the application being processed). About 95% of the time not defined as hands-on was attributed to work waiting in queue.

In addition to mapping the existing process and quantifying the workload, the project team performed a documentation of customer specifications and a documentation of design options. Questions about service were developed and circulated to gather broad input from agents. This survey was supplemented by a series of customer focus groups. The data collected by the team were categorized by four performance dimensions: quality, delivery, cycle time, and cost. From this research a set of process specifications was developed to guide the process redesign.

Designing the New Process

There were two stages to the design of the new process: high-level design and detailed design. In the initial high-level design, team members strove to think out of the box to create a conceptual model that would exceed the already ambitious design specifications. The driving specifications for the design were to dramatically improve responsiveness to customer needs and drastically reduce cycle time.

Sub-teams worked in parallel to design the new business. The best attributes of each sub-team were integrated into one cohesive vision of the new process and supporting structures. The resulting high-level design envisioned a work-team approach. Each work team would be aligned regionally with agents. A key feature of the high-level design concept was that it provided guidance for designing the details. The Time team created a new detailed business process map as the primary documentation of the new design. The new process contained only 85 process steps, more than 60% of which added value for the customer.

A New Organizational Structure

The new process called for new roles and responsibilities throughout the organization, where the traditional vertical organizational structure with first-line supervisors was replaced with a flatter organizational structure. Considerable responsibility was assumed by teams under this new organizational structure. Employees were then matched with the skills required in the newly identified roles and intensive training began. The entire division participated in planning and implementing the transition to the new process.

A new organization was to be structured around core teams that are regionally aligned with agents. There will also be a technical resource center that will be used to continually train people. Teams will pull resources from the technical resource center when trends indicate a higher volume of work, or if high enough, new teams will be formed.

Successful Results

The new design has resulted in significant process improvements, which have in turn had a substantial impact on Time. The process improvements have increased quality and delivery to agents while reducing Time's unit cost and cycle time. Revenue growth has been significant due to the exceptional service given to the agents. The increased flexibility with the new team structure gives the company a competitive advantage and also permits quick changes to regulatory constraints and provides Time with a solid base for transition in the new health care environment. Other results include:

- A 60% reduction in policy reissuances
- A 50% increase in measured customer satisfaction ratings
- A 10% reduction in cost per policy issued
- An 80% reduction in process cycle time for fast-track applications
- Significant increases in revenue from higher customer retention

QUESTIONS

1. Using the Deming Cycle, evaluate Time Insurance's process improvement efforts.
2. What are some other quality tools that Time Insurance could have used to better understand and improve its service levels?
3. What were the key success factors in this case, particularly as they relate to process redesign?
4. Discuss the relationship between process improvement and customer-added value.

Appendix B:

BUSINESS PROCESS ASSESSMENT TOOL

BUSINESS PROCESS ASSESSMENT TOOL*

Robert C. Preziosi

INTRODUCTION

Perhaps the most important issue facing organizations is productivity. Every major economy considers productivity growth a significant indicator of economic health. Changes in technology alone will not be sufficient fuel to spur productivity growth. The best opportunity for productivity improvement lies in business process improvement.

People say that things fall through the cracks. Time is often wasted as we wait for someone to forward something so that work can begin on a project. These are but two examples of ineffective business process. Whether the environment is manufacturing, service, or government, the potential possibilities for process improvement are already present in the organization. Improving business processes is fundamental to improving competitive position. However, before processes can be improved their levels of efficiency must be addressed.

ASSESSING BUSINESS PROCESSES

The Business Process Assessment Tool (BPAT) is designed for organizations and process improvement consultants. It is ideally suited for analyzing business processes to determine whether or not changes should be made. The BPAT will provide useful data for follow-up discussion and decision-making. The BPAT is an action research tool that could be used to identify process deficiencies and guide process improvement initiatives.

---

* Developed by Robert C. Preziosi, Professor of Management, School of Business, Nova Southeastern University. © 2000. All rights reserved.

Four elements are used in this tool to assess business process improvement: (1) organization leadership and policy; (2) employee practices; (3) customer needs; and (4) supplier perspective. Each element has separate standards that are analyzed by a series of questions. A positive assessment results when all standards are met.

Organization leadership and policy includes those standards that are broad and require the focused activity of those leaders held most responsible for business process improvement. *Employee* practices refers to those actions employees must take or be the target of for value-adding business process improvement. *Customer needs* includes those special areas where business process improvement directly influences customer relationships. *Supplier perspective* refers to those considerable value-adding actions of special interest to suppliers.

BUSINESS PROCESS ASSESSMENT TOOL

Directions: Please respond to each of the following statements about your organization using the scoring system below. Please give your assessment of your organization by placing the number of your response in the blank space next to the statement.

1 Occurs all the time
2 Occurs some of the time
3 Occurs rarely
4 Occurs never

Organization Leadership and Policy

___1. The culture of this organization is supportive of business process improvements.

___2. I apply my knowledge base to improve what I do.

___3. This organization applies its knowledge base to improve processes.

___4. All opportunities to improve processes in this organization are acted upon.

___5. A cost-benefit analysis is conducted before moving forward with a process improvement.

___6. Measurement and performance standards are used to improve processes.

___7. The result or output that a business process is supposed to produce is clearly stated.

___8. Business processes that fail to enhance the attainment of business objectives are eliminated.

___9. We have approaches in place which clearly measure input resource utilization of all business processes.

___10. We have a precise knowledge of where, when, and how every input enters a business process.

___11. All departments use business process improvements to add value in our organization.

___12. A reporting system is used that measures variation from business process standards.

Employee Practices

___1. All employees are trained in the methods of business process improvement.

___2. Every business process has a champion who owns the process and continuously seeks to improve it.

___3. Training is always provided when a business process is improved.

___4. Employees consider technological solutions when deciding ways to improve a business process.

___5. Employees are trained in the use of "blueprinting" of our process activities via process flow diagrams.

___6. Employees are trained in methods of working together in teams across organization functions.

___7. Employees seek process improvement solutions that keep identified problems from recurring.

___8. Employees want the best value-adding process in place.

___9. Employees are properly rewarded for business process improvements which they produce themselves.

___10. Managers and supervisors are trained in coaching skills to help employees with their business process improvement projects.

Customer Needs

___1. We evaluate changing customer requirements to determine if they are consistent with our mission and values.

___2. Customers are involved in improving business processes that directly affect them.

___3. We communicate to our customers internal staff changes that affect how and with whom they will conduct business with our organization in the future.

___4. Business process improvements that will negatively impact customers are avoided.

___5. Data from customers are used when considering a process improvement opportunity.

___6. Training on our improved business processes is provided to customers so that they are kept up to date on our business processes that impact their business.

___7. Business process improvements translate into enhanced customer service.

___8. We have a very clear plan with our customers on when, how, and where feedback about business processes will be given.

___9. Our organization only implements business process improvements that have a positive impact on customer service.

___10. Our organization has a database that will tell us if a customer is lost because of poor business process(es).

Supplier Perspective

___1. Suppliers are kept informed of our business process improvements.
___2. We respond to suppliers to improve our business processes.
___3. We only conduct business with suppliers who believe in ongoing business process improvement.
___4. We require that suppliers inform us whenever they change a business process that directly affects us.
___5. The organization's supply chain is well understood by all employees.
___6. What our suppliers provide to our organization is consistently monitored for quality.

Scoring

Add up the scores for each element and total those four scores.

___   Organization Leadership and Policy
___   Employee Practices
___   Customer Needs
___   Supplier Perspective
___   TOTAL

As a general guideline, the total score indicates the following:

133–152	Superior business processes
114–132	Excellent business processes
95–113	Acceptable business processes
0–94	Unacceptable

Note: The best diagnosis results from conducting a question-by-question analysis.

NOTES

[1] Harrington, H. G., Hoffherr, G. D. and Reig, Jr., R. P. (1999). *Area Activity Analysis.* New York: McGraw-Hill.

[2] Losyk, B. and Preziosi, R. C. (1998). *Customer Service Audit.* Davie, FL: Innovative Training Solutions.

[3] Preziosi, R. C. and Ward, P. J. (1998) *Strategic Target Actions Review, in the 1998 Annual,* Vol. II, Consulting, San Francisco: Jossey–Bass/Pfeiffer.

[4] Weinstein, A. and Johnson, W. J. (1999). *Designing and Delivering Superior Customer Value,* Boca Raton, FL: St. Lucie Press.

Appendix C:

FINAL SURVEY QUESTIONS AND DETAILED CORRELATION AND REGRESSION RESULTS

SUPPLY CHAIN DECISION PROCESS ASSESSMENT*

Supply Chain Management

Definition: the process of developing decisions and taking actions to direct the activities of people within the supply chain toward common objectives.

The purpose of this survey is to capture the current status of your decision activities necessary for the successful operation of your supply chain. This survey attempts to capture YOUR OPINION concerning what is done and how often, who does it, and how it is done.

Thank you for your participation.

> ### Decision Process Area: PLAN
> Includes P1: Plan Supply Chain, and P0: Plan Infrastructure

Please circle your answers concerning this supply chain
decision process area using a range of:

1–never or does not exist, 2–sometimes, 3–frequently, 4–mostly, 5–always
or definitely exists
Please put an "X" on any question you are unable to answer.

1. Do you have an operations strategy planning
   team designated?................................   1 2 3 4 5

2. Does this team have formal meetings?..............   1 2 3 4 5

3. Are the major supply chain functions (sales, marketing,
   manufacturing, logistics, etc.) represented on this
   team? 1 2 3 4 5

4. Do you have a documented (written description,
   flow charts, etc.) operations strategy planning
   process? 1 2 3 4 5

5. Is there an owner for the supply chain
   planning process? 1 2 3 4 5

6. Has the business defined customer priorities? 1 2 3 4 5

7. Has the business defined product priorities? 1 2 3 4 5

8. When you meet, do you make adjustments
   in the strategy and document them? 1 2 3 4 5

9. Does the team have supply chain performance
   measures established? 1 2 3 4 5

10. Does the team look at the impact of its strategies
    on supply chain performance measures?.............   1 2 3 4 5

11. Does the team use adequate analysis tools to
    examine the impact before a decision is made?.......   1 2 3 4 5

12. Is the team involved in the selection of supply
    chain management team members?...................   1 2 3 4 5

13. Does this team look at customer profitability? 1 2 3 4 5

14. Does this team look at product profitability? 1 2 3 4 5

15. Does this team participate in customer and
supplier relationships?. 1 2 3 4 5

16. Do you analyze the variability of demand
for your products?. 1 2 3 4 5

17. Do you have a documented demand forecasting
process? 1 2 3 4 5

18. Do your information systems currently support
the Demand Management process? 1 2 3 4 5

19. Does this process use historical data in developing
the forecast? 1 2 3 4 5

20. Do you use mathematical methods (statistics)
for demand forecasting? 1 2 3 4 5

21. Does this process occur on a regular
(scheduled) basis?. 1 2 3 4 5

22. Is a forecast developed for each product?. 1 2 3 4 5

23. Is a forecast developed for each customer?. 1 2 3 4 5

24. Is there an owner for the demand management
process? 1 2 3 4 5

25. Does your demand management process make use
of customer information?. 1 2 3 4 5

26. Is the forecast updated weekly? 1 2 3 4 5

27. Is the forecast credible or believable? 1 2 3 4 5

28. Is the forecast used to develop plans and
make commitments? 1 2 3 4 5

29. Is forecast accuracy measured? 1 2 3 4 5

30. Are your demand management and production
planning processes integrated? 1 2 3 4 5

31. Do sales, manufacturing and distribution organizations
collaborate in developing the forecast? 1 2 3 4 5

32. Overall, this decision process area performs
very well. 1 2 3 4 5

Decision Process Area: SOURCE

Includes P2: Plan Source

Please circle your answers concerning this supply chain
decision process using a range of:

1–never or does not exist, 2–sometimes, 3–frequently, 4–mostly, 5–always
or definitely exists
Please put an "X" on any question you are unable to answer.

1. Is your procurement process documented (written
   description, flow charts)? 1 2 3 4 5

2. Does your information system support this process? ...   1 2 3 4 5

3. Are the supplier inter-relationships (variability, metrics)
   understood and documented? 1 2 3 4 5

4. Is a process owner identified? 1 2 3 4 5

5. Do you have strategic suppliers for all products
   and services? 1 2 3 4 5

6. Do suppliers manage "your" inventory of supplies? 1 2 3 4 5

7. Do you have electronic ordering capabilities
   with your suppliers? 1 2 3 4 5

8. Do you share planning and scheduling information
   with suppliers? 1 2 3 4 5

9. Do key suppliers have employees on your site(s)? 1 2 3 4 5

10. Do you collaborate with your suppliers to develop
    a plan? 1 2 3 4 5

11. Do you measure and feedback supplier
    performance? 1 2 3 4 5

12. Is there a procurement process team designated? 1 2 3 4 5

13. Does this team meet on a regular basis? 1 2 3 4 5

14. Do other functions (manufacturing, sales, etc.) work
    closely with the procurement process team members? ..   1 2 3 4 5

15. Overall, this decision process area performs
    very well. 1 2 3 4 5

Decision Process Area   MAKE

Includes P3: Plan Make

Please circle your answers concerning this supply chain
decision process using a range of:

1–never or does not exist, 2–sometimes, 3–frequently, 4–mostly, 5–always
or definitely exists
Please put an "X" on any question you are unable to answer.

1. Do you have a documented (written description, flow
   charts, etc.) production planning and scheduling
   process? 1 2 3 4 5

2. Are your planning processes integrated and
   coordinated across divisions? 1 2 3 4 5

3. Do you have someone who owns the process?. 1 2 3 4 5

4. Do you have weekly planning cycles?. 1 2 3 4 5

5. Are supplier lead times a major consideration
   in the planning process? 1 2 3 4 5

6. Are supplier lead times updated monthly?. 1 2 3 4 5

7. Are you using constraint-based planning
   methodologies? 1 2 3 4 5

8. Is shop floor scheduling integrated with the overall
   scheduling process?. 1 2 3 4 5

9. Do your information systems currently
   support the process? 1 2 3 4 5

10. Do you measure "adherence to plan"?. 1 2 3 4 5

11. Does your current process adequately address the
    needs of the business? 1 2 3 4 5

12. Do the sales, manufacturing, and distribution organizations
    collaborate in the planning and scheduling process? . . .   1 2 3 4 5

13. Is your customer's planning and scheduling
    information included in yours? 1 2 3 4 5

14. Are changes approved through a formal, documented approval process? 1 2 3 4 5

15. Are plans developed at the item level of detail? 1 2 3 4 5

16. Overall, this decision process performs very well...... 1 2 3 4 5

Decision Process Area: DELIVER

Includes P4: Plan Deliver

Please circle your answers concerning this supply chain
decision process using a range of:

1–never or does not exist, 2–sometimes, 3–frequently, 4–mostly, 5–always
or definitely exists
Please put an "X" on any question you are unable to answer.

1. Is your order commitment process documented
   (written description, flow charts, etc.)? 1 2 3 4 5

2. Do you have a promise delivery (order commitment)
   "process owner"? 1 2 3 4 5

3. Do you track the percentage of completed customer
   orders delivered on time? 1 2 3 4 5

4. Are the customers satisfied with the current on-time
   delivery performance? 1 2 3 4 5

5. Do you meet short-term customer demands from
   finished goods inventory? 1 2 3 4 5

6. Do you "build to order"? 1 2 3 4 5

7. Do you measure customer "requests" versus actual
   delivery? 1 2 3 4 5

8. Given a potential customer order, can you commit
   to a *firm* quantity and delivery date (based on actual
   conditions) on request? 1 2 3 4 5

9. Are the projected delivery commitments given to
   customers credible (from the customer's view)? 1 2 3 4 5

10. Do you promise orders beyond what can be satisfied
    by current inventory levels? 1 2 3 4 5

11. Do you maintain the capability to respond
    to unplanned, drop-in orders? 1 2 3 4 5

12. Do you automatically replenish
    a customer's inventory? 1 2 3 4 5

13. Do the sales, manufacturing, distribution, and planning organizations collaborate in the order commitment process? 1 2 3 4 5

14. Do your information systems currently support the order commitment process? 1 2 3 4 5

15. Do you measures out-of-stock situations? 1 2 3 4 5

16. Is your order commitment process integrated with your other supply chain decision processes? 1 2 3 4 5

17. Is your distribution management process documented (written description, flow charts, etc.)? 1 2 3 4 5

18. Does your information system support distribution management? 1 2 3 4 5

19. Are the network inter-relationships (variability, metrics) understood and documented? 1 2 3 4 5

20. Is a process owner identified? 1 2 3 4 5

21. Are impacts of changes examined in enough detail before the changes are made? 1 2 3 4 5

22. Are changes made in response to the loudest "screams"? 1 2 3 4 5

23. Are deliveries expedited (manually bypassing the normal process)? 1 2 3 4 5

24. Do you use a mathematical tool to assist in distribution planning? 1 2 3 4 5

25. Can rapid replanning be done to respond to changes? 1 2 3 4 5

26. Is the distribution management process integrated with the other supply chain decision processes (production planning and scheduling, demand management, etc.)? 1 2 3 4 5

27. Does each node in the distribution network have inventory measures and controls? 1 2 3 4 5

28. Do you use automatic replenishment
    in the distribution network? 1 2 3 4 5

29. Are process measures in place? 1 2 3 4 5

30. Are they used to recognize and reward
    the process participants? 1 2 3 4 5

31. Overall, this decision process area performs very well . . 1 2 3 4 5

> ## Common Themes within Each Supply Chain Decision Process Area
> Strategies, tactics and philosophy components that are common across the supply chain

Please circle your answers to the following questions in regards to your opinion of the OVERALL supply chain.

1. Your supply chain processes are documented and defined

Not at all	a little	somewhat	mostly	completely
1	2	3	4	5

2. Your supply chain organizational structure can be described as

Traditional Function Based	a little Process	some Process	mostly Process	entirely Process Based
1	2	3	4	5

3. Your supply chain performance measures can be described as

Traditional Function Based	a little Process	some Process	mostly Process	entirely Process Based
1	2	3	4	5

4. People in the supply chain organization can be generally described as

Totally Internally Focused	a little Customer Focused	somewhat Customer Focused	mostly Customer Focused	entirely Customer Focused
1	2	3	4	5

5. Your information systems currently support the supply chain processes

Not at all	a little	somewhat	mostly	completely
1	2	3	4	5

6. The demand for your product varies

Not at all	a little	somewhat	often	always
1	2	3	4	5

7. Jobs in the supply chain can generally be described as

Limited Task Oriented	a little Process	somewhat Process	mostly Process	Broad Process Oriented
1	2	3	4	5

Relative Performance

1. Please rate the overall performance of your business unit last year.

Poor	Fair	Good	Very Good	Excellent
1	2	3	4	5

2. Please rate the overall performance of your business unit last year relative to major competitors.

Poor	Fair	Good	Very Good	Excellent
1	2	3	4	5

3. Compared with your major competitors your overall inventory days of supply (DOS) are:

Poor	Fair	Good	Very Good	Excellent
1	2	3	4	5

4. Compared with your major competitors your overall cash-to-cash cycle times are:

Poor	Fair	Good	Very Good	Excellent
1	2	3	4	5

5. Compared with your major competitors your delivery performance vs. commit date is:

Poor	Fair	Good	Very Good	Excellent
1	2	3	4	5

6. Compared with your major competitors your quoted order lead times are:

Poor	Fair	Good	Very Good	Excellent
1	2	3	4	5

General Questions Needed for Analysis and Reporting of Results

Please circle your answers to the following questions.

1. What is your industry ?

a) Electronics	e) Aerospace & Defense	i) Pharmaceuticals/Medical
b) Transportation	f) Chemicals	j) Mills
c) Industrial Products	g) Apparel	k) Semiconductor
d) Food & Beverage/CPG	h) Utilities	l) Other_____

2. Within what function do you work?

a) Sales	e) Manufacturing	i) Purchasing
b) Information Systems	f) Engineering	j) Other_____
c) Planning & Scheduling	g) Finance	
d) Marketing	h) Distribution	

3. What is your position in the organization?

a) Senior Leadership/Executive

b) Senior Manager

c) Manager

d) Individual Contributor

Contact Information (Optional):
Name _____
Title _____
Company _____
Address _____
City/State/Zip _____
Phone _____ Fax _____
E-mail _____

BUSINESS PROCESS ORIENTATION QUESTIONNAIRE*

The purpose of the attached survey is to gather data for a study investigating the relationship between Business Process Orientation and organizational performance.

Thank you for your participation in this survey.

The following questions ask you to comment on your organization. What we wish to know is how you perceive the way your organization is structured toward getting work done. Each question will ask you to agree or disagree on the following scale.

(PLEASE CIRCLE <u>ONLY ONE</u> NUMBER FOR EACH QUESTION)

Completely Disagree	Mostly Disagree	Neither Agree Nor Disagree	Mostly Agree	Completely Agree	Cannot Judge
1	2	3	4	5	8

---

* This survey is the property of Kevin McCormack and cannot be duplicated or used without permission.

Process View (PV)

(PLEASE CIRCLE <u>ONLY ONE</u> NUMBER FOR EACH QUESTION)

Completely Disagree 1	Mostly Disagree 2	Neither Agree Nor Disagree 3	Mostly Agree 4	Completely Agree 5	Cannot Judge 8

1. The average employee views the business as a series of linked processes. 1 2 3 4 5 8

2. Process terms such as *input, output, process,* and *process owners* are used in conversation in the organization. 1 2 3 4 5 8

3. Processes within the organization are defined and documented using inputs and outputs to and from our customers. 1 2 3 4 5 8

4. The business processes are sufficiently defined so that most people in the organization know how they work. 1 2 3 4 5 8

Process Jobs (PJ)

1. Jobs are usually multidimensional and not just simple tasks 1 2 3 4 5 8

2. Jobs include frequent problem solving. 1 2 3 4 5 8

3. People are constantly learning new things on the job. 1 2 3 4 5 8

Process Management and Measurement Systems (PM)

1. Process performance is measured in the organization. 1 2 3 4 5 8

2. Process measurements are defined. 1 2 3 4 5 8

3. Resources are allocated based on process. 1 2 3 4 5 8

4. Specific process performance goals are in place. 1 2 3 4 5 8

5. Process outcomes are measured. 1 2 3 4 5 8

Interdepartmental Dynamics (ID)

(PLEASE CIRCLE <u>ONLY ONE</u> NUMBER FOR EACH QUESTION)

Completely Disagree 1	*Mostly Disagree* 2	*Neither Agree Nor Disagree* 3	*Mostly Agree* 4	*Completely Agree* 5	*Cannot Judge* 8

Interdepartmental Conflict

1. Most departments in this business get along well
   with each other.   1 2 3 4 5 8

2. When members of several departments
   get together, tensions frequently run high.   1 2 3 4 5 8

3. People in one department generally dislike
   interacting with those from other departments.   1 2 3 4 5 8

4. Employees from different departments feel
   that the goals of their respective departments
   are in harmony with each other.   1 2 3 4 5 8

5. Protecting one's departmental turf is considered
   to be a way of life in this business unit.   1 2 3 4 5 8

6. The objectives pursued by the marketing
   department are incompatible with those of the
   manufacturing department.   1 2 3 4 5 8

7. There is little or no interdepartmental conflict
   in this business unit.   1 2 3 4 5 8

Interdepartmental Connectedness

1. In this business unit, it is easy to talk with virtually anyone you need to, regardless of rank or position.

1 2 3 4 5 8

2. There is ample opportunity for informal "hall talk" among individuals from different departments in this business unit.

1 2 3 4 5 8

3. In this business unit, employees from different departments feel comfortable calling each other when the need arises.

1 2 3 4 5 8

4. Managers here discourage employees from discussing work-related matters with those who are not their immediate superiors or subordinates.

1 2 3 4 5 8

5. People around here are quite accessible to those in other departments.

1 2 3 4 5 8

6. Communications from one department to another are expected to be routed through "proper channels."

1 2 3 4 5 8

7. Junior managers in one department can easily schedule meetings with junior managers in other departments.

1 2 3 4 5 8

Organizational Performance (OP)

(PLEASE CIRCLE <u>ONLY ONE</u> NUMBER FOR EACH QUESTION)

Completely Disagree	Mostly Disagree	Neither Agree Nor Disagree	Mostly Agree	Completely Agree	Cannot Judge
1	2	3	4	5	8

Measures of Esprit de Corps

1. People in this business unit are genuinely concerned about the needs and problems of each other.   1 2 3 4 5 8

2. A team spirit pervades all ranks in this business unit   1 2 3 4 5 8

3. Working for this business unit is like being part of a family.   1 2 3 4 5 8

4. People in this business unit feel emotionally attached to each other.   1 2 3 4 5 8

5. People in this business unit feel as if they are "in it together."   1 2 3 4 5 8

6. This business unit lacks an esprit de corps.   1 2 3 4 5 8

7. People in this business unit view themselves as independent individuals who have to tolerate others around them.   1 2 3 4 5 8

Overall Performance (1 = poor; 5 = excellent)

1. Please rate the overall performance of the business unit last year.   1 2 3 4 5

2. Please rate the overall performance of the business unit last yearrelative to major competitors.   1 2 3 4 5

General Questions Needed for Analysis and Reporting of Results

Please circle your answers to the following questions.

1. What is your industry?

1. Electronics	5. Aerospace & Defense	9. Pharmaceuticals/Medical
2. Transportation	6. Chemicals	10. Mills
3. Industrial Products	7. Apparel	11. Semiconductor
4. Food & Beverage/CPG	8. Utilities	12. Other_____

2. What is the approximate size of your entire company (number of employees)?

Small <1,000 _____ Medium 1,000 – 10,000 _____ Large >10,000 _____

3. Within what function do you work?

1. Sales	5. Manufacturing	9. Purchasing
2. Information Systems	6. Engineering	10. Other_____
3. Planning & Scheduling	7. Finance	
4. Marketing	8. Distribution	

4. What is your position in the organization?

1. Senior Leadership/Executive

2. Senior Manager

3. Manager

4. Individual Contributor

Contact Information (Optional):
Name_____
Title_____
Company _____
Address_____
City/State/Zip_____
Phone _____   Fax_____
E-mail _____

DETAILED CORRELATION AND REGRESSION RESULTS

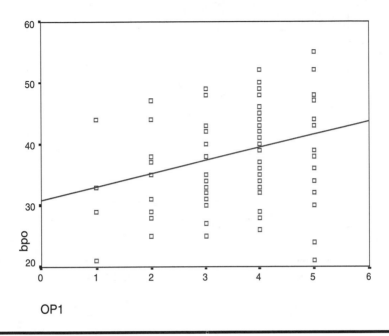

Figure C.1 Regression Line — BPO vs. Overall Performance (OP)

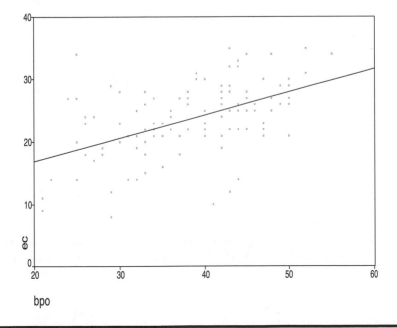

Figure C.2 Regression Line — BPO vs. Esprit de Corps (EC)

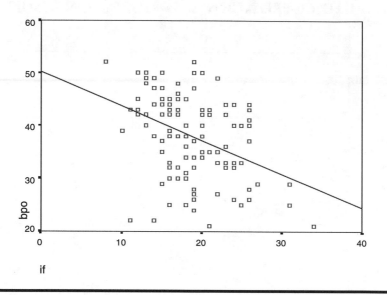

Figure C.3 Regression Line — BPO vs. Interfunctional Conflict (IF)

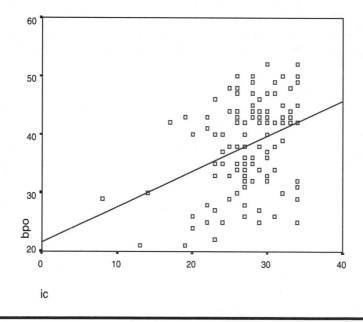

Figure C.4 Regression Line — BPO vs. Interfunctional Connectedness (IC)

Table C.1 BPO Standardized Regression (Beta) Coefficients

	BPO	EC	IC	ID	IF	OP1
BPO	1.000	.5005	.3658	.5219	−.3800	.2792
	(114)	(113)	(112)	(113)	(113)	(111)
	P = ?	P = .000	P = .000	P = .000	P = .000	P = .003

Table C.2 Correlation Matrix Results — Independent and Dependent Variables

Variables	Factor 1 — PM	Factor 2 — PJ	Factor 3 — PV
Factor 1 — PM	1.000	0.183**	0.507*
Factor 2 — PJ	0.183*	1.000	0.278**
Factor 3 — PV	0.507**	0.278**	1.000
Dependent — ID			
Conflict — IF	−0.325*	−0.231*	−0.279**
Connectedness — IC	0.309*	0.262**	0.187**
Validity — OP1	0.319*	0.206*	0.111***
Validity — EC	0.428*	0.313*	0.308*

*Significant at the 0.01 level. ** Significant at the 0.05 level. ***Significant at P = 0.248.

GLOSSARY

Benchmarking: The systematic comparison of process performance, practices, and attributes for the purpose of process improvement.

Business process: A collection of activities that takes one or more kinds of input and creates an output that is of value to the customer. A reengineered business is composed of strategic, customer-focused processes that start with the customer and emphasize outcome, not mechanisms.

Business Process Change: A strategy-driven organizational initiative to improve and (re)design business processes to achieve competitive advantage in performance through changes in the relationships among management, information, technology, organizational structure, and people.

Business Process Orientation (BPO): Emphasizes process, a process oriented way of thinking, customers, and outcomes as opposed to hierarchies.

Coordination theory: A body of principles about how activities can be coordinated and how actors can work together harmoniously.

Collaboration: Forms, behaviors, constructive conflict, and creative integration.

Core Processes: The value-added activities that support and facilitate the customer life cycle, representing the foundation of most businesses, the value that customers pay for, and the essence of most businesses.

Enabling Processes: Processes that are key to the achievement of critical business goals such as online order processing that enablers an Internet retailer to exist.

Esprit de corps: The feeling of belonging to a group and the strong identification with the group goals and purpose.

Horizontal corporation: Described as eliminating both hierarchy and functional boundaries. It is governed by a skeleton group of senior executives that include finance and human resources. Everyone else is working together in multidisciplinary teams that perform core processes, such as product development, with only three or four layers of management between the chairman and the "staffers" in a given process.

Interfunctional coordination: The coordinated utilization of company resources to create superior value for target customers.

Interdepartmental dynamics: Consists of *conflict* and *connectedness*. Conflict pertains to the extent to which the goals of different departments were incompatible and tension prevailed in interdepartmental interactions. Connectedness captures the extent to which individuals in a department were networked to various levels of the hierarchy in other departments.

Intra-organizational collaboration: Among people and across units.

Kaizen: The overriding concept behind good management; a combination of philosophy, strategy, organization methods, and tools needed to compete successfully today and in the future.

Marketing Cycle: The key marketing functions performed by goods- and service-producing organizations, including but not limited to distribution, sales, logistics, pricing, customer service management, and promotion.

Organizational culture: The pattern of shared values and beliefs that helps individuals understand organizational functioning and thus provides them with the norms for behavior in the organization.

Process: A specific ordering of work activities across time and place, with a beginning, an end, and clearly identified inputs and outputs as a structure for action.

Process centering: Refocusing and reorganizing around processes or building an organization with a business process orientation.

Process Flow Diagram: Tool used for defining the steps of a process in order to better understand the importance and value of each step to the customer and identify potential fail points.

Process management: Viewing the operation as a set of interrelated work tasks with prescribed inputs and outputs. Provides a structure and framework for understanding the process and relationships and for applying the process-oriented tools. Establishing control points, performing measurements of appropriate parameters that describe the process, and taking corrective action on process deviations.

Process Maturity Model: A model depicting increasing levels of process performance.

Process-oriented structure: An organization structure that de-emphasizes the functional structure of business and emphasizes the process; cross-functional view. A dynamic view of how an organization delivers value.

Process View: The cross-functional, horizontal picture of business involving elements of structure, focus, measurement, ownership, and customers.

Reengineering: The development of a customer-focused, strategic business process-based organization enabled by rethinking the assumptions in a process-oriented way and utilizing information technology as a key enabler.

Supply Chain: The global network used to deliver products and services from raw materials to the end customer through engineered flows of information, physical distribution, and cash.

Supply Chain Management: The process of developing decisions and taking actions to direct the activities of people within the supply chain toward common objectives.

Supply Chain Networks: Groups of supply chains that are voluntarily connected and cooperating for the purpose of serving a specific market or set of customers.

Supporting processes: Not as insignificant, as their position on the map might imply. They are shown at the bottom of the map because they are the furthest from the customer. Human resource management would be an example of a supporting process for a consulting company. Information systems frequently serve as a supporting process for many companies today.

Sustaining processes: Critical to the operation of the business, but may not result in direct customer interactions, e.g., product research and development.

Teams: Groups of individuals who work together to develop products or deliver services for which they are mutually accountable.

Value: A trade-off between the benefits received and the costs (both economic and noneconomic) incurred in purchasing and using a product or service.

Value chain: A systematic way of examining all the activities a firm performs and how they interact to provide competitive advantage (see Figure 2.3). This chain is composed of "strategically relevant activities" that create value for a firm's customers.

Value Proposition: A "shared" understanding between the firm and customers or an implicit "contract" between company and customer, listing all products, programs, services, and target customer, and the effect of these offerings on the customer's business.

Vertical organization: An organization whose members look up to bosses instead of out to customers. Loyalty and commitment are given to functional fiefdoms, not the overall corporation and its goals. Too many layers of management still slow decision-making and lead to high coordination costs.

INDEX